Changing Pathways

Changing Pathways

Forest Degradation and the Batek of Pahang, Malaysia

Lye Tuck-Po

LEXINGTON BOOKS
Lanham • Boulder • New York • Toronto • Oxford

LEXINGTON BOOKS

Published in the United States of America
by Lexington Books
An imprint of The Rowman & Littlefield Publishing Group, Inc.
4501 Forbes Boulevard, Suite 200, Lanham, Maryland 20706

PO Box 317
Oxford
OX2 9RU, UK

British Library Cataloguing in Publication Information Available

Library of Congress Cataloging-in-Publication Data

Lye, Tuck-Po.
  Changing pathways : forest degradation and the Batek of Pahang, Malaysia / Lye
Tuck-Po.
       p. cm.
  Includes bibliographical references and index.
  ISBN 0-7391-0650-3 (cloth : alk. paper)
  1. Batek (Malaysian people)—Social conditions. 2. Batek (Malaysian people)—
Government relations. 3. Batek (Malaysian people)—Politics and government. 4.
Indigenous peoples—Ecology—Malaysia—Pahang. 5. Forest ecology--Malaysia--
Pahang. 6. Forest degradation--Malaysia--Pahang. 7. Forest conservation—Malaysia—
Pahang. 8. Pahang—Social conditions. 9. Pahang—Environmental conditions. I. Title.
DT595.2.B38L94 2004
304.2'8'095951—dc22

                                                          2004013922
Printed in the United States of America

To the memory of my parents,
Lye Toong Lee and Chung Ling Yoke

Contents

Illustrations

Figures

Photos

Tables

Maps

Abbreviations and Acronyms

a.s.l.   above sea level
DWNP   Department of Wildlife and National Parks; formerly the Game Department
ha   hectares (10,000 square meters; 2.471 acres)
JHEOA   Jabatan Hal-Ehwal Orang Asli (Department of Orang Asli Affairs); formerly the Department of Aborigines
K.   Kuala (Malay for "confluence")
Kg.   Kampung (Malay for "village")
km   kilometer
lit.   literally
m   meter
Sg.   Sungai (Malay for "river")
UMNO   United Malays National Organisation, the dominant political party in the ruling coalition

Spelling and Writing Conventions

Words in Batek are transcribed in this volume according to the orthography currently used in Mon-Khmer linguistic studies. The symbols employed are pronounced approximately as indicated below:

Symbols	Pronunciation	Examples
i	as the *ee* in English *bee*, but shorter	*labiʔ* (river turtle)
e	as the *ai* in English *bait*	*teʔ* (land; earth)
ɛ	as the *e* in English *bet*	*kabɛn* (kinfolk; friends)
ɨ	approximately as the *oo* in many varieties of English *good* or the *u* in Scottish *hus* (house)	*cip* (to walk)
ə	as the *e* in English *the*	*bəlaw* (blowpipe)
a	as the *u* in English *fun*	*kap* (to bite or sting)
u	as the *u* in English *bull*	Gubar (thunder-god)
o	as the *oa* in English *boat*	*jok* (to move)
ɔ	as the *au* in English *taut*, but shorter	*kiyɔm* (under; below)
c	as the *ch* in English *change*	*can* (foot)
ɲ	as the *ny* in English *banyan*	*ɲoʔ* (to tell a lie)
ŋ	as the *ng* in English *singer* (not as in *finger*)	*baŋkol* (eaglewood)
ʔ	glottal stop, as the *tt* in London Cockney *little*	*naʔ* (mother)
k	as the *k* in English *kite* (never as a glottal stop)	*kəpok* (orphans)

Nasal vowels are written with a superscript tilde: *lõʔ* (to wash into the river)

Grammar and word formations

Batek grammar and word-formation conventions have not been fully studied yet. Recognized word modifiers that appear in this book include: *bah-* (locator, usually suggesting directionality); *pi-* (causative marker); *pə-* (causative marker); *kə-* (relative marker; possibly also a locator); *tə-* (relative marker); *-m* (intention marker); *bə-* (progressive, modifying both nouns and verbs).

All but the intention marker are separated from the modified word by a hyphen. For example: *kə-sɛn* (that is past); *tə-bəw* ([specified subject] that is big); *bah-kiyɔm* (to-below, i.e., downriver or underside); *pi-dɛŋ* (to cause to see, i.e., to point); *pə-hinɛ̃k* (lit. to cause to imitate: i.e., to show how something is said or done); *hehm* (you-all and we-all want); *bə-kəpok* ("to be orphaning"). Also written in the same way are reduplicative verbalization of nouns (for example, *kiy-kuy* [to be heading] from *kuy* [head]) and verb distributives (for example, *bit-bɛ̃t* [to hang on tightly here and there, i.e., to be entwined] from *bɛ̃t* [to hang on tightly]). Derivations are written as single words (for example, *bəlahɔt* [term of abuse for *takop* yams] from *bəlhɔt* [watery tubers]).

I have slightly simplified the spellings of some personal names. For example, the name ʔalɔr appears here as Alɔr; Gʔ appears as Gk. Relationship terms prefixed to teknonyms (see "Notes on Names") are treated as words rather than names and not capitalized except for when they start a sentence. For example, ʔeyGk; naʔAlɔr; taʔKadɔy; yaʔKaw.

The glossary (pp. 201–9) contains notes on the most frequently used words and particles in this book.

Sorting order. Phonemes in the glossary and index are sorted as follows: *a, ã, b, c, d, e, ɛ, ɛ̃, ə, ə̃, g, h, i, i̶, j, k, l, m, n, ɲ, ŋ, o, õ, ɔ, p, r, s, t, u, w, y, ʔ.*

Notes on Names

As the Batek say, only after the creator beings had named the primordial plants, animals, and landscape features could we know them. Many other cosmologies and religions around the world feature the naming process as a necessary business of creation. Likewise, anthropologists are fond of names: of the people we work with, the places we work in, the things and relationships that the people consider nameworthy. Of course we could put too much emphasis on language. Many practices, beliefs, environmental features, and relationships are never named (and some names are so dangerous that to utter them is to invite chaos). And many groups of people never had names until there were people who needed names to call, codify, or colonize them by. Still, giving names and labels is the easiest way to show (among many things) "*This* is what I'm talking about, not *that*." So some explanation of naming conventions is in order here.

For personal names, I follow Batek conventions: personal names for children, nicknames for childless adults, and teknonyms for other adults. Teknonyms, as translated from the Batek words for them, mean "children's names" or "grandchildren's names." They are easily recognized through one of these prefixes (all relationship terms): ?ey (father), na? (mother), ya? (grandmother), and ta? (grandfather). So when an adult male is named (for example) ?eyTəltil, that means "father of Təltil" and Təltil is probably his oldest unmarried son or daughter. A teknonym with the prefix "ya?" or "ta?" should therefore signal that the person in question is an adult of at least (by Batek standards) middle-age.

There is one exception: Tebu, who is central to this book. He should have a teknonym, since he is already a grandfather, and he does. However, I expect that readers of this book will not find it easy to remember all the teknonyms, and Tebu's name must be easily remembered. So I have given him a Malay pseudonym (it means "sugarcane"), which is also widely used by indigenous communities in Malaysia.

For place-names, where there is an official name and a Batek name for a location (a toponym) or river (a hydronym), as far as possible the Batek name is given first, followed by the official name in parentheses, such as "Was Tahaɲ (Malay[1] Kuala Tahan)." Unless the official name is more appropriate for the discussion context, subsequent references will be to the Batek name. All official names follow the current spellings and versions: for example, Taman Negara rather than King George V National Park (the founding name that lasted until Independence in 1957), even when the context of discussion is colonial.

As for the Batek's group name (their "ethnonym"), I do not use dialect group labels here. Following the Batek's usual practice, I identify subgroups by the state where they generally live: Pahang, Kelantan, or Terengganu (see map 4).

In the linguistic and some anthropological literature, the Batek's language is commonly identified as Batek De'. The dialect group classifications were first announced by Kirk Endicott, who distinguished several subgroups that were associated with defined geographical areas: Batek De' (the numerically dominant group), Batek Nòng, Mendriq, Batek Teh, Batek 'Iga,' and Batek Te'. Recently, he has argued more forcefully that dialect groups do not hold constant through time—which means that there are no true "dialects" or dialect speakers.[2] I agree with this newer position. Further, the subgroup names are too ephemeral to be meaningful. I found that people were often puzzled when asked to identify who belongs in which dialect group, a line of questioning that inevitably ended with the firm sendoff "Everyone is Batek" (I cannot say this for the Mendriq, as I have never met them). As far as I can tell, only two dialect group labels are widely recognized by the Pahang Batek: Batek 'Iga' and Batek Te'. It is more natural for the Batek to identify people by hydronyms. Thus, the Batek Te' are better known as *batɛk Kɔh* ("Koh people," after the Koh River valley in Kelantan where they lived). Hardly anyone that I've met will confess to being the Batek De', though their language belongs in that category! However, it is possible that dialect group labels have more meaning in Kelantan (where Endicott focused his studies) and Terengganu; the amnesia described earlier seems to lift—a little bit—when there are visitors from these other states.

There are two other "Batek" groups in Pahang who are usually lumped with them (this lumping tendency means that official census numbers for the "Batek" category are not reliable). These are the Batek Nòng and the Batek Tanum. The Batek do not consider their languages intelligible with theirs. They know of, but have almost no interaction with the Nòng people, who do not even neighbor them geographically. So this group is almost certainly not a "dialect group" of the Batek but an entirely separate population of people, who took up agriculture some time in the 1920s or so.

The Tanum people were identified as Mintil by Geoffrey Benjamin,[3] a name that those I've met fully reject. They are enclaved by Malays on one side (especially in the Tanum River valley from which they take their name) and the Batek on the other (along the Kechau River) and therefore no longer have the land they need to maintain a fully mobile way of life (see maps 2 and 4). Today they are found mainly in three small villages, which receive government attention: Sg. Garam near the Malay village of Dada Kering (most visited by the Batek and known by them and the Tanum as Marɛm), Paya Keladi, and Telok Gunong in Cegar Perah on the Temetong River near Jerantut town. The combined population total was 175 in 1995 government records (but I have some reason to think this number an underestimation). One clue to their customary adaptation is the old folk Malay name for them, which is now remembered in Batek as *batɛk bɔlukɛr* (secondary forest people). There is some intermarriage between them and the Batek and what appears to be an expanding socioeconomic domain.

I have never met the Nòng but have stayed on and off with the Tanum for a few days each time. I collected an approximately 600-word Marɛm Tanum vocabulary (unpublished), but little socioeconomic information of any scholarly value. As such, these prefatory remarks are practically all that I will say in this book about either group. Unfortunately, no contemporary linguist or anthropologist has done systematic, long-term fieldwork with either the Nòng or Tanum peoples (although there is enough early documentation to give a reasonably adequate impression of life in pre-World War II days).[4] On present evidence, I recommend that Batek Nòng and Batek Tanum be considered as distinct from the Batek and therefore classifiable as two additional ethnic minorities in the Peninsula.

Notes

1. See chapters 1 and 5 for further discussion of Malay influences on Batek language and culture.
2. Kirk Endicott 1997.
3. Benjamin 1976.
4. The main references are indexed in Lye 2001.

Acknowledgments

The first words of the manuscript that became my doctoral dissertation flashed into my head while I was still at fieldwork—one afternoon as I sat in a long wooden boat powered with an outboard engine, rounding a bend of the Tembeling River in Pahang, Malaysia. Months later in Honolulu, with fieldwork over and a computer screen in front of me, I was surprised that those words still rang true. Soon enough, the dissertation was done.

I then took various drafts of this manuscript from Hawai'i back to Malaysia, then to Japan and back again to Malaysia. I worked on it as time permitted and brought portions of it into the wide-open spaces of seminars and conferences around the world. The years went by, and this book came to look less and less like its forebear. As my focus was pulled away by an undertow of stresses and crises, I comforted myself with the conceit that I was doing exactly as a true hunter-gatherer should: seek my knowledge and inspiration on the move, wherever the opportunity presents itself. The ride has not been smooth; if not for the friends, companions, and mentors who joined me on this long journey, I would never have arrived. There are many people and institutions that I must thank.

To start, I'd like to thank those who have passed on: from my grandmother Ng Foong Lin and my father Lye Toong Lee to, among the Batek, those whose names appear in chapter 8 (pp. 167–68). I truly regret that they did not live to see this book. I am intensely sustained by the memory of my grandmother, to whom I owe the greatest debt of all. My earliest lessons in kinship and social roles and responsibilities were learnt literally at her feet. My father, who sank into a long silence when I announced my intention to study anthropology, did not know how much he himself had started it all, when he gave me my first stamp collection and stirred my curiosity of places and peoples far far away. I am grateful that he always intuited the force of my ideals and never, ever, pulled me from my path.

The people who most shaped the route taken by this book are of course the Batek. It seems terribly unfair to mention some names and not others and I will not do so. Though not everyone was equally involved in my studies, collectively they hosted, tolerated, fed, taught, and worried over me. If a year passes and I do not go back to the forest, I know I *will* hear about it next time. They opened their homes, made their time available, and shared my sorrows. They are kin, friends, and mentors. Those families who let me live in their lean-tos and houses need special mention. They willingly endured my frequent and sometimes exasperating requests for information and left me alone when I needed time to think, sleep, or write. Certainly they will also remember the many times they protected my fieldnotes and kept the children from bothering me!

What sparked my first professional interest in the Batek was Kirk Endicott's monograph on their religion. I am grateful beyond measure to Kirk for his invaluable correspondence, sharing his insights on the Batek, and responding to my work. He and his wife Karen Lampell Endicott have given us a pioneering record and set high standards of excellence in research and writing; they have raised challenging questions that forced me to consider more nuanced interpretations. Christian Vogt finished his dissertation fieldwork with the Batek as I arrived for mine and offered important insights. Stepping on the heels of other fieldworkers has for me been one of the great joys of doing research.

Most of this book is based on data collected during the 1993 and 1995–1996 fieldwork seasons and, as such, I must thank those teachers who supervised the early florescence of my ideas. Michael R. Dove and Alice Dewey were the two mainstays of my years at the University of Hawai'i, Manoa. Michael, despite being snowed under by his new responsibilities in New Haven, continues to read and comment on my work and he remains a source of great inspiration. During our sojourn in Hawai'i, he spent much time pulling me out of one scrape or mood after another. I am indebted to him for all his help. Alice long ago provoked me to question some of the conceptual frameworks that I too lazily pinned my ideas on. In the last three years, the main supervision of my work was taken over by P. Bion Griffin—my committee chair, teacher, and friend. Bion's enthusiasm for hunter-gatherer comparisons and willingness to spend hours in chat and counsel had a galvanizing effect, made the dissertation-writing process seem much more fun than it probably was, and gave me a necessary ethnological perspective on my fieldwork data. Even now, he continues to let out a loud and much-appreciated email guffaw at the latest comic anecdote from the field.

Lots more people have taught, inspired, encouraged, provoked, or helped me in one way or another. I wish I could name them all: but I'd never be able to finish such a list. I thank Gordon P. Means for sharing insights from the unpublished notes of Paul Means and, further, Saw Leng Guan at the Forest Research Institute of Malaysia for his botanical identifications. For teaching me everything I know about Aslian linguistics, I'll always be grateful to Geoffrey Benjamin, Niclas Burenhult, and Gérard Diffloth. Long chats with Sonja Brodt, Geoffrey Davison, Deanna Donovan, Edmond Dounias, Mitsuo Ichikawa, Jenne Lajuni, Jean Kennedy, Jane Mogina, Ong Hean Chooi, Raj Puri, and Dennis Yong helped to unravel the mysteries of fieldwork and forest ecologies.

Among teachers, I hope I do no one a disfavor by singling out Carolyn Bell, Robert K. Dentan, Jon Goss, Alan Howard, Terry Rambo, and Leslie Sponsel. Not all have taught me formally but all have suffered through enough of my prose to deserve my gratitude. For their warm hospitality, during which some of this manuscript was written, I'd like to thank Alex and Carolyn Bell, Lee Swee Fun, David Loh, and Mano Maniam. Mano read drafts of manuscript chapters, as did Donna Amoroso, Niclas Burenhult, Bob Dentan, Michael Dove, David Frossard, Tom Gill, Bion Griffin, Kirk Endicott, Rosemary Gianno, and Colin Nicholas. Their feedback has been indispensable.

For making possible the conditions for writing this book, I am especially grateful to Koji Tanaka for inviting me to take up a Japan Society for Promotion of Science (JSPS) postdoctoral fellowship (1998–2000). Research and writing were further enabled by Visiting Fellowships in 2001: at the National Museum of Ethnology in Osaka (to which I returned in 2002 and 2003) and the Resource Management in Asia-Pacific project of the RSPAS, the Australian National University. For their kind invitations to these institutions, I thank, respectively, Ken-ichi Abe and Fadzilah M. Cooke.

By the time I put the finishing touches to this manuscript, I was enjoying a fruitful year at Randolph-Macon Woman's College in Virginia. I am grateful to the college for inviting me to take up the Quillian Visiting International Professorship; most of all, I thank the students in my courses on tropical forest ethnographies. Their curious and insightful questions about the Batek were always provocative, and I hope they will continue to ask questions.

Portions of this book were first prepared for workshops and seminars in Kyoto, Kuching, Los Baños, Perth, and Halle. For inviting me to these venues, which gave me the chance to explore the wider relevance of Batek ideas, I would like to thank, in order, Katsuyoshi Fukui, Michael Leigh, Michael R. Dove and Percy S. Sajise, David Trigger, and Thomas Widlok. This book has also benefited from comments received during presentations at Dartmouth College and the Universiti Kebangsaan Malaysia, Bangi, and I am indebted to Kirk Endicott and Sumit Mandal for inviting me there.

Fieldwork in 1995–1996 was funded by the East-West Center and the Wenner-Gren Foundation for Anthropological Research. The John D. and Catherine T. MacArthur Foundation provided additional support through the project, "The Conditions of Biodiversity Maintenance in Asia." For making possible that assistance and the further support from the project, "The Institutional Context of Biodiversity Conservation in Southeast Asia" (1999), I am indebted, once again, to Michael R. Dove and Percy S. Sajise. Later field visits were enabled by the JSPS postdoctoral fellowship and I must thank, as ever, Koji Tanaka.

I also thank the Economic Planning Unit in the Prime Minister's Department, Kuala Lumpur, for granting research permission; the State Economic Planning Unit in Kuantan for permitting me to work in Pahang; and the Wildlife Department for permitting me to travel in and out of the national park, Taman Negara. Within this last agency, I especially thank staff members at the Kuala Tahan park headquarters. Hood M. Salleh and the Department of Sociology and Anthropology at Universiti Kebangsaan Malaysia, Bangi, were my local sponsors in 1995–1996. I am most grateful to Hood, foremost for his role in facilitating the granting of my research permit. His company is well-appreciated too.

Finally, there is one person whose help has been indispensable. This is the Batek's rattan trader, Towkay Mang, who keeps me informed about the Batek's current locations and goes out of his way to ferry me to and out of Batek camps whenever I ask. I have had relatively trouble-free field experiences, thanks in large part to the kindness of traders like him, and of those anonymous individuals who responded to my thumbs for rides. On the Tembeling River, local villagers, park rangers, and tour guides readily boated me around. I am most grateful for these acts of random kindness—invariably offered without any question—that made it so much easier to move around.

Portions of chapters 3, 5, and 7 were originally published in the journal *Southeast Asian Studies* as "Forest, Bateks, and degradation: Environmental representations in a changing world" (volume 38, 2000, pp. 165–84) and "The significance of forest to the emergence of Batek knowledge in Pahang, Malaysia" (volume 40, 2002, pp. 3–21) and are reproduced here with permission of the Center for Southeast Asian Studies, Kyoto University.

[handwritten annotations:]

new kind of ethno

> not the dutiful collection of details, the state not in sight, no cultural isolates

> neither the endangered "bravery libraries" or salvage ethnog

— both a peon + an intervention

Friction → industrial Logging

Bratils → Penan.

1

Introduction *Kirsch → PNG.*

On June 3, 1993, I began to record stories from a small group of Batek men and women on the Kechau River in north central Pahang state, Peninsular Malaysia. Strong winds had blown up late that afternoon. Thunder rumbled; rain pelted down. Some people ran out to the foliage-clear logging road beside the camp. They urged me to do the same. If any of those wind-tossed trees fell over the camp, we would be prime targets.

A felicitous moment to start asking questions. That night, after the winds had calmed, the rain lessened, and everyone returned to their lean-tos, I asked a group to tell me stories: stories about the thunder-god, Gubar; about animals and, then, as I put it in Malay, *apa-apa pun boleh* (lit. anything also-can). It wasn't long before they introduced the myths of origin, of fire and rivers, the world, and ethnic groups. Half-a-dozen people, two women and four men of various ages, sat by the fire. Others sat in, walked out, put in a voice or response here and there. I daresay still others were listening from their lean-tos. Two or three hours later, as one story led to another and one idea connected with another, they started to tell me about the problems of losing the forest.

Well into the night, I heard about Gubar the thunder-god: not the mischief-maker as portrayed in some of the popular tales, but a fearsome being whose anger can cause violent destruction. In this persona, he represented something else—the moral order. The messages from the shamans, I heard, amounted to this:

People—the townspeople, the royalty, and government officials—must believe the Batek. If people do not believe in their stories and continue to open up the forest for development purposes, to carve out roads, to turn natural forest into monocrop plantation estates, there will be no forest left. Gubar will be angry, the *ethnographer, researched project as a "messenger" to powerful th. + outside "courier" (p. 17)*

floods will come, the land will dissolve, the underground snake (a kind of Earth deity) will surface, and the world will collapse.

"Our stories," said one man and his wife to silent agreement from the others, "you may regard as *surat* [letters] that we give to you that you may deliver them to the *orang bandar* [Malay: townfolks]."[1] This couple was ʔeyTow and naʔTow, who had appointed themselves my "parents."

People don't believe Batek stories, he continued. "Luckily, I've a daughter who's come here to ask me for our stories. If people don't believe, I don't know what we'll do. I think we'd surely die."

"If they don't believe, let them come here. I don't have money to go to them," naʔTow's father ʔeyBajaw added.

These comments, and the setting in which they were gathered, provide an apt introduction to the major concerns of this book. This is an anthropological account in the sense that it contributes to the literature on the Batek, who are little known even in Malaysia, and offers a certain perspective on their perceptions, norms, activities, and modes of behavior. I hope, however, that this is more than another record of, in popular parlance, "a vanishing way of life." This way of life is changing and always has changed; following a common pattern in Peninsular Malaysia, the Batek's ideas and concepts, social relations, economic activities, and even their language and stories show evidence of long-term openness to influences from outside. I am not worried about cultural survival issues. Those will take their course and the Batek will deal with them as they have always dealt with socio-cultural change. My major concerns are the material conditions that enable and constrain the Batek's capacity to pursue their chosen modes of life, conditions that are vastly contemporary and of global relevance: the degradation of the forests from which the Batek take their identity. My approach, then, is environmental.

I focus on the implications of degradation for the Batek and for the world that is enveloped in the notion of the *surat* (letters), which I examine in more detail below. The political intent of my study is to further along the process of communication between the Batek and readers of this book. Like the Batek's stories, this book can be considered a *surat*. Before elaborating on my concerns and the origins of this book, I present below an introduction to the Batek and to the general issues of social change and environmental degradation in Malaysia.

The Batek, the Environment, and Change

The Batek call themselves batɛk həp (people of the forest). They are among the score or so indigenous ethnic minorities of Peninsular Malaysia, the Orang Asli (Malay: original people; see map 2), and live in lowland tropical forests in the states of Pahang, Kelantan, and Terengganu (see map 4). I've worked only in Pahang and, within that, Kecau and Tembeling, which are subdistricts of Lipis and Jerantut districts (see maps 3, 7, and 8).

Plate 1. Moving camp: a halt along the journey down the Tenor River, Taman Negara, January 1996.

The most established town in this area is undoubtedly Kuala Lipis (Batek Was Lipis), strategically located at the confluence of the Jelai and Kechau rivers and, until the rise of Jerantut town in recent decades, the main gateway into the interior of Pahang. These hinterlands were long renowned for their wealth of forest products and minerals (especially gold). Lipis district's importance in the state economy declined as coastal and maritime travel quickened pace in the nineteenth century, and it remains among the least built-up districts of this largely rural Malaysian state. At 35,965 square km, Pahang is the largest state in the Peninsula (and the third largest after Sabah and Sarawak in East Malaysia [Borneo]), containing the tallest mountain (Gunung Tahan) and the longest river (Pahang River). And, Geoffrey Benjamin remarks, it exhibits the greatest indigenous linguistic diversity of any Malaysian state.[2]

Once entirely covered in forest (mainly lowland dipterocarp), it remains one of only four states where forest cover exceeds 50 percent of the land area. One factor hindering forest clearance in the past was the remoteness of its interior. Thus in the 1920s and '30s when plans were mooted to protect forest within the future Taman Negara national park (which straddles parts of Kecau and Tembeling districts), the state government was willing to surrender land because it saw no clear commercial interests there (for more on Taman Negara's founding, see chapter 5, pp. 106–7). Until quite recently movement in Pahang was almost entirely by water, with the Pahang River and its major tributaries being the standard highways. Indeed, population was sparse and settlements tended to hug the banks of these rivers, thus leaving the interior to those who knew how to work their way around it: Orang Asli. Today, population density remains one of the lowest among the states.

The critical natural barrier to land development is the Main Range, formed of ancient sedimentary rocks. Malaysians commonly depict this in metaphors: a "spine" or "backbone" dividing the Peninsula into two distinct halves, the east and west coasts. Pahang occupies a somewhat anomalous position in this division, its northern regions sheltered by the Main Range but its southern half largely comprising deltaic plains that could be reached from neighboring states. For example, in the fifteenth and sixteenth centuries the Penarikan trade route led from the east coast, through Pahang River and the upper reaches of the Muar (in present-day Negri Sembilan), and thence to Melaka (Malacca) on the west coast, for a long time the major commercial and religious center in the Peninsula. Thus, though travel was arduous and lengthy and the forest inhibited settlement, Pahang was never isolated geographically.

The Main Range did hinder the building of east-west land links. The East Coast railway line connecting Kelantan state on the northeast with the populous west coast was not completed till 1931; this line runs through Lipis and Jerantut districts, serving many rural communities, and remains the line most familiar to the Batek (see map 3). In 1995–1996 when I occasionally traveled to fieldwork on this train, the travel time from Kuala Lumpur (the capital city) to Kuala Lipis was something like eight to nine hours. This is a considerable investment of time for most west coast Malaysians, for whom the interior of states like Pahang continues to seem quite remote. My own father traveled around Pahang in the 1960s and, he recalled, it seemed like miles upon miles of "nothing." Surfaced roads first appeared in the Malayan landscape in the decade after the introduction of the car in 1902, but initially road networks were fragmentary and, as Ooi puts it aptly, were designed to serve the river traffic; i.e., to take people to the rivers where the "real" traveling was done.[3] Even up till 1958 the road maps show a blank space for the interior of Pahang; only one main road connected Kuala Lipis with the state capital on the east coast, Kuantan.

Another important mountain range is the Tahan on the northeast corner where Pahang abuts Kelantan and Terengganu. In Gunung Tahan, it rises to an elevation of 2187 m. Today the mountain has achieved fame as a climbing destination but it has always been an orographic barrier to human settlement. There is another ancient trade route along this range; this route starts from the coast up the Tembeling (Batek Təmiliŋ) and thence the Tahan (Batek Tahaɲ) Rivers, from there crossing the mountains and down into Kelantan on the other side. Today perhaps only the Batek have an intimate knowledge of the various passes and passages through these mountains, but they tend to shun the truly rugged, high-altitude areas. The watersheds of all the major Batek rivers in the three states are in this mountain range.

The borders of the Batek territory in Pahang are the two rivers that give their names to the subdistricts: Kəciw (Malay Kechau) on the west and Təmiliŋ on the east. The Kəciw rises in the Tahan mountain range and feeds into the Jelai, which in turn meets the Təmiliŋ at Kuala Tembeling. The Təmiliŋ in turn is a tributary of the Pahang River. Innumerable rivers and streams drain into the Kəciw and Təmiliŋ; the hills and valleys of those tributaries are where I have done most of my work (see maps 5 to 8).

Government authorities, following (if unself-consciously) the classification system first codified by Paul Schebesta in 1926, recognize three ethnolinguistic groupings among the Orang Asli: Semang, Senoi, and Aboriginal Malay. These names are also used by anthropologists, whose criteria for classification is slightly different. The Batek are categorized as Semang, which refers to a cluster of classically mobile hunting-and-gathering societies living mainly in the northern and north-central parts of the Peninsula and the bordering areas of Thailand. Semang are commonly known by the racial term "Negrito" but as this Spanish word means nothing more than "little black people," it is best dropped altogether. Even as a racial term, "Negrito" disguises the high degree of physical heterogeneity among Semang populations. Semang and Senoi peoples speak varieties of Aslian, the name given by linguists for the Mon-Khmer languages that are found in the Peninsula (Mon-Khmer is a division of the Austroasiatic language family). In current linguistic opinion, the Aslian history in the Peninsula is at least three millennia old. Semang are speakers of the North Aslian sub-branch, which includes the languages of Batek, Batek Tanum, Batek Nòng, Kensiw, Kintaq, Jahai, Mendriq, and the various Meniq peoples of South Thailand. Aboriginal Malays (or Melayu Asli) peoples are speakers of the Malayo-Polynesian branch of the Austronesian family of languages, which includes Malay, the official language in Malaysia.[4]

The total Batek population is not determined but is unlikely to exceed seven to nine hundred, i.e., roughly 0.73 percent of all Orang Asli or less than 0.004 percent of all Malaysian citizens in the 2000 census.[5] As of 2001, the Pahang population is around 450. Most of the other Semang peoples have to some degree or another become sedentarized, although that sometimes means that they use settlements like base camps, relocating periodically for a range of social and economic reasons. Sedentarization (the cessation of mobility) is pronounced among Batek in Kelantan and Terengganu; following the general model, the Kelantan Batek were reported (as of 1990) to spend up to three months a year in the forest collecting rattan and other forest products.[6] Quite a number of Kelantan groups migrated to Pahang at different times in the past twenty years, mainly to flee environmental destruction. Not much is known of the Terengganu Batek, other than that some have intermarried and co-reside with Semaq Beri in Ulu Trengganu villages. In view of the general trend, the Batek of Pahang are the only statewide Semang population who remains as full-time mobile hunter-gatherers. By mobile hunting and gathering is meant that way of life so often chastised by government officials as "nomadic." We will encounter these criticisms again. For now, a straightforward description of the Batek way of life will suffice.

The Batek's mobility is not "nomadic" in the sense of being utterly free of ties to the land. Indeed, as we will see in this book, if the Batek did not have ties to the land, they could not be mobile. One cannot just wander randomly around the forest; it is much too complex a landscape for that. Without topographic and resource knowledge to start with, it is not possible to be mobile. The development of that knowledge over generations fosters important bonds and sentiments: both among people who share a place, and between people and the land. Contrary to popular perception, hunter-gatherers tend *not* to be expansionist. They do not habitually move into other people's territories unless it makes absolute sense: land loss, displacement, outmigration of neighboring populations, and government resettlement are among the usual reasons. For one thing, they rarely have the military muscle to just take over a territory. They depend rather on social networking, either among themselves or with friendly dominant groups, to extend the spatial resources available to them. This seems to be a common theme around the world.

Practically, the traditional mode of dwelling in the forest is to live in a camp (*hayã?*); these camps are connected by an extensive series of pathways that traverses over walking trails, rivers, and logging roads, in a topography marked by the alternation of land and water. The Batek's preferred ecological niche is the forested foothills where altitude rarely exceeds 200 m above sea level (a.s.l.). The forest formations are classified as lowland dipterocarp (<300 m a.s.l.) and hill dipterocarp (300–700 m a.s.l.) forests, which together form the lowland evergreen rain forest. Here there is a complex interlacing of feeder streams and tributaries and the terrain is characterized by rolling hills, humpback ridges, ridge crests, river meanders, rounded capes, and escarpments. In other words, despite the absence of a rugged surface, this environment contains many distinctive landforms and is not featureless. The Batek do not ascend to the higher elevations that lack food sources and they demonstrate no knowledge of montane ecology.

Batek social-economic strategies, and their concern over environmental degradation, must be placed within the broader context of change in Malaysia. Driven by export-led growth, general aspiration is towards development, with industrialization being the objective. In the 1980s up until the Asian financial crisis of 1997, this was characterized by high growth rates. As promulgated in the 1991 National Development Policy, the government is targeting 2020 as the year for achieving fully developed status. The growth of urbanization (due to natural increase, net migration, and reclassification of urban areas) is one indicator of just what's happening all around. Just over a quarter of the population was classified as "urban" in 1970; thirty years later, that number has exceeded 60 percent. Consumption of fossil fuels (oil, gas, coal) is alarmingly high and there appear to be no government efforts to systematically disengage this from economic growth. Income levels (and, correspondingly, material consumption) have gone up as well, from RM6099 per capita in 1990 to RM14788 per capita in 2000 (at the current exchange rate, from USD1605 to USD3891.58). General statistics of this sort can be misleading. As Gomez and Jomo summarize: "although the reforms instituted by the New Economic Policy (NEP) reduced poverty substantially and led to the growth of ethnic Malay middle and business classes, there has been growing concern over the influence of political patronage on the business sector, the increasing inequitable distribution of wealth, and the apparent increase in corruption and other abuses of power."[7] A disproportionate number of Orang Asli rank among the officially defined "hardcore poor." The development model is very much one of "trickle down," with a slant towards development of "growth poles" (urban nodes) and social-economic services there (including health care and education) being superior to those in less densely populated rural areas. Poverty eradication is a major objective of social development programs.

Historically, development has been fueled by the liquidation of natural assets: timber, minerals, and offshore oil and gas reserves. The trend has been towards increasing openness and globalization of the economy. Until the mid-1980s, rubber, oil palm, timber, and tin were the main income generators. In the mid-1980s, due to the fall in commodity prices, there was a structural transformation in the economy with alternative sectors, notably industrialization and tourism, assuming more importance. By 1997, the "agriculture, forestry, livestock, and fishing" sector stood at 12.2 percent of the GDP (down from 30.8 percent in 1970), while "manufacturing"—focused on export-oriented production since 1980—was at 35.5 percent (up from 13.4 percent in 1970).[8] Overall, sustainability concerns tend to be sidelined.

> The twenty-first century will not be able to proceed like the last one. . . . [T]here are too few natural renewable resources remaining to sustain the excessive exploitation rates of the previous century. . . . [T]he steady output of greenhouse gases, toxic chemicals, hazardous waste and even organic loads into the environment overwhelms the natural capacity of forests, oceans, land and the atmosphere to absorb pollutants and contaminants. Nature also has its limits and Malaysia is on the verge of testing that hypothesis to the extreme.[9]

The process has long been unfolding, but as far as contemporary Batek forests are concerned the artificial "jumping off" point would be the large scale rural development projects that began in the 1960s and 1970s.

A key feature has been forest conversion and clearance for a range of purposes. Initially, the main culprits were probably the establishment of extensive plantation estates and the roads and infrastructure to link up rural areas with market centers. (There has been a parallel history of smallholder plantations; the environmental impact of these has not been as great and therefore will not concern us here.) The history of large-scale plantation estates goes back to the end of the nineteenth century, when rubber was introduced as the main export crop. That has now been overtaken by oil palm; Malaysia has cornered the market in palm oil. In the early 1970s, forest clearance began to make inroads into hitherto remote areas, including much of the Batek territories in interior Kelantan. Latterly, with the shift towards manufacturing and industrialization, the built environment has expanded dramatically, with more and more forest lost to property development, industrial parks, and infrastructure like airports, ports, highways, and hydropower projects. Since the adoption of tourism as a key revenue earner in the 1980s, this has included rapid construction of tourist resorts and other recreational facilities (like golf courses), even in ecologically fragile areas like hillsides and, as we'll see, protected areas. All these continue or are in planning and have led to the irreversible fragmentation of the forest landscape. There are important implications for the configuration of space:

> Land is increasingly valued for its site and locational attributes rather than its inherent ecological properties. Land-use changes accelerated by leap-frogging urban-industrial impositions into rural areas and the appreciation in the value of adjoining agricultural land at almost exponential rates further diminish the significance of ecological attributes of land. Land is invariably held for speculative purposes and to build up the "land bank" of corporate bodies awaiting further high value-added transformations at opportune moments.[10]

To put all this in simple terms, *the city is moving into the village while the forest is retreating.* The traditional divisions between rural and urban, forest and village, are increasingly breaking down as are local values of land (especially those concerning rights of tenure, access, use, and property distribution) while new divisions have appeared—notably, increasing disparities of wealth both within indigenous communities and in broader Malaysian society generally. There are effects on soils, waters, air, biological species, and, less discussed, human psychology in the face of rapid change. For forest-dwelling Orang Asli, most of whom do not have title to land, these translate into the loss of subsistence bases, social-political impacts aside. Land security now ranks high on any list of the Orang Asli's pressing needs.

Internationally, Malaysia has made a commitment to keep at least 50 percent of the total land area under permanent forest cover. However, the actual extent of land remaining under forest cover is fiercely disputed, with government agencies using a variety of criteria—depending on what the statistic is being used for—to determine what constitutes "forest." These criteria may not correspond to their scientific or indigenous counterparts. As Sahabat Alam Malaysia (an environmental group) points out: "Sorting out the statistics on forest cover in Malaysia is a difficult task. Depending on the source, whether it be the Department of Forestry, Ministry of Primary Industries (MPI) or the Ministry of Science, Technology and the Environment (MOSTE), each has its own methodology and reasoning for presenting the numbers." One fundamental problem is in how these various agencies determine what category a block of forest should be placed in: "data presentation from different agencies is neither uniform nor consistent, thus leading to discrepancies and conflicting analyses."[11]

A Forest Research Institute of Malaysia (FRIM) report on the effects of climate change on the forestry sector (presumably drawing on official sources) shows what can be achieved by playing with categories. In 1994, the size of "forested land" was put at 58.61 percent of the total land area (in the Peninsula, 44.1 percent of the land area) but this included plantation forests (comprising fast-growing afforestation species, chiefly eucalyptus, pine, and *Acacia mangium*). When "agriculture tree crops" (oil palm, rubber, cocoa, coconut) were included, "total tree cover" went up to 72 percent of the land area![12] This is close to the high number that government officials will sometimes cite as "land under forest cover" but at least 15 percent of it is composed of planted, often alien, species, which come nowhere near to reproducing the natural habitats and ecosystems of the tropical forests.

Geographers Teh et al., drawing on a variety of sources and official forestry statistics, note that between 1947 and 1995, the annual deforestation rate was 0.9 percent or a loss of 156,250 ha per year (defining deforestation as "the complete clearing of tree formations and their replacement by some other use of land").[13] According to the Department of Wildlife and National Parks (DWNP) of the federal-level Ministry of Science, Technology, and the Environment, the period 1966 to the 1980s saw a loss of forested areas in the Peninsula of some 25 percent. Analysing land use data for 1990, the DWNP concluded that only 47 percent of the Peninsula's land area was covered in forest by their criteria (note the discrepancy with the FRIM report cited above). Most of the loss is in the lowland dipterocarp forest; the DWNP recognizes that those at altitudes of less than a hundred meters are among the key threatened habitats in the country.[14] Pahang is regularly recognized among the states where most of the forest loss in this period has occurred. (It is also regularly identified among the states that exceed their annual allowable cuts of timber.)

In Sahabat Alam Malaysia's summary: "By the judicious use of the term 'tree cover,' which includes man-made forests of tree crops, around 75% of the country's total land area is said to be covered in leafy goodness. In reality, natural forests cover less than 60% of the landscape, but in Peninsular Malaysia the percentage dwindles to less than 45%."[15] Even these revised statistics might be too high for some. For example, according to Teh et al. "the forest cover is now 67.2%, 59.8% and 41.5% in Sarawak, Sabah and Peninsular Malaysia respectively."[16] Whatever the actual numbers involved, the unquestionable trend is that "as natural forests are harvested for wood products, tree crops and plantations increase to keep the green meter running at acceptable levels."[17] For forest peoples, this trend promotes "de-territorialization" in more ways than one. Plantations of the sort favored by governments are high input, high capital investment monocultural systems that are humanly uninhabitable. As forests (both primary and secondary) shrink, not only do indigenous claimants lose land area, the range of ecosystem services that they derive for everyday needs (food, water, biomass for fuel and construction, soil protection and conservation, arable land, etc.) will decline both in quantity and quality.

The general summary is that there are few lowland dipterocarp forests left outside the protected area system. Remaining in good forest condition are timber reserves (officially, "production forest"), game, bird, and marine sanctuaries, state parks, and totally protected areas (national parks). Ultimately, many of these are remnants; with growing landscape fragmentation they are becoming "forest islands" that may not be extensive enough to sustain the original biodiversity. Enrichment planting of secondary and logged-over forests is, though promoted, poorly implemented.

The oldest and, in the Peninsula, largest totally protected area is Taman Negara (4343 sq. km; 1677 sq. miles), the national park founded in 1938/1939, which mostly sits astride Batek territory in Pahang, Kelantan, and Terengganu. The park is run by the DWNP, which maintains park headquarters at Was Tahaɲ (Malay Kuala Tahan). The Tembeling River forms the southeastern border of the park as well as the traditional boundary between Batek and Semaq Beri territories. Historically, the primary goal of protected areas is the protection of wildlife habitats; however, in the past twenty years this ideal has been somewhat subordinated to tourism imperatives. Taman Negara remains the largest unbroken tract of forest available to the Batek, who are permitted to live there and travel in and out of the park at their will, but not to collect <u>forest products, including fauna, for sale</u>. They are regarded as the original inhabitants of the park but do not have an administrative role and are not consulted on management issues.

Park borders are vulnerable to commercial development and encroachment. On the Terengganu side some parts of the park are flooded within the catchment area of the Kenyir hydroelectric dam and new dams are in planning in Kelantan. The highway to Gua Musang (a town in Kelantan) slices through a corner of the western part in Pahang (see map 3). Much of the fringe forests outside the park are in various logged-over conditions. In some areas, these forests are converted to monocrop plantations (primarily of oil palm) while in others they are left to regenerate. The ecological ideal of maintaining unbroken stretches of forest corridor for the free passage of wildlife is probably elusive in some places. The Batek make extensive use of the fringe forests. They will return to logged-over forests so long as standing trees remain and food resources are available.

Other responses to environmental change include:

- they are less mobile, meaning that they stay longer in a place;
- there is more sharing of space, so people are not as widely dispersed as before and camp group populations might also be higher than under conditions of land abundance;
- they incorporate logged-over areas into their territorial network—which, of course, may have been part of their territory all along;
- they buy more of their food than before, thus maintaining a fairly even impact on the ecology;
- they are more conscious of territorial boundaries like those of Taman Negara; and
- some groups tend to spend more time in semi-permanent settlements.

There are two settlements in my fieldwork area, Was ?ato? (Malay Kuala Atok) and Was Yɔŋ (Malay Kuala Yong), located an hour's boat ride from each other and considered by the Batek to lie in distinct ecological zones and territories (see map 5). Though there is much contact and family groups may move between them, as communities they are distinct and the people do not want to be relocated (in Malaysian administrative idiom, "regrouped") as one. Was ?ato? is a government-provided settlement situated just outside park borders. Was Yɔŋ, located inside the park, was spontaneously established by the Batek after they were told to leave their original settlement at Was Tahaɲ (a ten-minute boat ride away). I am intimate with Was Yɔŋ and its habitual residents but hardly know Was ʔatoʔ.

At any point in time, there may be some ten different groups dispersed throughout the forest. The population in a camp (called here a "camp group" following Kirk and Karen Endicott) averages 36.2 persons, though if several groups are co-residing, this number might be tripled (for details, see Table 1.1). Flux is an ever-present part of life. This is well described by Kirk Endicott: "The composition of a group might change daily, as some families left and new ones joined it."[18] I would add that youths, especially the young men, are most likely to be changing camps frequently and restlessly. Some of this flux is shown in Table 1.2, which details average camp sizes and deviations from the norm. It shows that only two camp groups underwent no population changes. The norm is that within the life cycle of a camp, the population will fluctuate and sometimes quite greatly (for example, going from forty-nine to seventy-two in one day alone). As people *jok* (travel from one place to another), they bring information with them and take information along. This is how groups keep in touch with each other's news, from gossip (like who's done what or who's given birth) to location details and intended movements. Flux therefore serves the larger purpose of enabling people to schedule movements, plan itineraries, and communicate with each other.

Social relations in the camps are integrated enough that each camp group should be considered a distinct community. There is a misleading impression that a camp is "just" a "temporary camp" as though it has no social meaning beyond its physical structure. Indeed, each camp is a village unto itself. Within the camp, the dwelling is a palm-thatched lean-to (also called *hayãʔ*) mounted on at least five posts; these days, plastic sheets (tarpaulins) are used, either as a substitute for the thatch or as an additional covering (see Plates 6 and 7). The most common type of domestic group living in a lean-to is the nuclear household, which might be a *kəmam*, a basic family unit consisting of parents and children, or a *kəlamin*, a childless couple. Usually there are around eight such lean-tos in a camp, with the remaining lean-tos housing aged (often widowed) elders or single-sexed groups of unmarried males and females (including young teenagers who have moved out of their parents' lean-tos).

A group might average two weeks in a camp; the distance between successive camps is from five to ten km, or one to two hours' walk. After three or four months, roughly corresponding to the end of a season, camp groups disband and splinter groups may move to other river valleys, joining and forming groups anew. Movement is predominantly on foot but longer journeys (like from one river valley to another) might involve hitching rides from traders. Membership in a camp is open to all but people tend to travel with and to kinfolk (*kabɛn*). Never in my experience has someone joined a camp where he or she did not have any prior ties, however attenuated.

Economically, hunting and gathering have been the production base but the Batek undertake a variety of other activities too. There are daily, seasonal, and annual changes in production activities. The core of the economy remains subsistence-based hunting and gathering. This does not mean that quantitatively

Table 1.1. Population statistics for Batek camps

	All camps (1993, 1995, 1996)
Mean population	36.20
Median	35
Mode	24
Standard deviation	14.73
Minimum population	6
Maximum population	72
Total days counted	226

(1) Data only for camps where continuous population changes were recorded

Table 1.2. Average camp populations, 1993, 1995–1996[1]

Camps (month/year)	Mean population	SD[2]	Range[3]	Total days counted
Pagar Sesak (6/93)[4]	20.64	7.05	6–29	11
Kəciw (7/93)	20.00	0	no change	15
Felda 6 (8/93)[5]	25.18	1.4	24–27	11
Kəciw (7/95–8/95)	61.38	5.27	50–67	29
Yuŋ–ʔatoʔ (8/95)	41.00	4.07	37–50	8
Tərŋin (9/95)	31.71	4.48	26–43	14
Tabɛn (9/95)	20.58	6.1	10–26	12
Rəmpay (10/95)	21.50	3.21	17–25	6
Taŋɔy (12/95–1/96)	38.29	2	34–44	21
Buməkəl (1/96)	33.89	0.33	33–34	9
Tabɛn (3/96)	22.22	12.81	13–42	9
Cərah (7/96)	23.44	1.67	19–24	9
Ruwiw camps (8/96–10/96):				
Pənacik	21.00	0	no change	4
Pacew	27.40	7.89	21–36	5
downriver Pacew	36.40	3.46	31–43	14
Wɛ̃c Sok Bawac	40.54	1.76	36–43	13
Was Ruwiw	33.55	7.38	28–48	11
Was Tiaŋ	47.00	8.66	42–57	3
Was Yɔŋ (10/96–11/96)	53.43	8.7	49–72	21
All camps	36.20	14.73	6–72	226

(1) Data only for camps where continuous population changes were recorded
(2) Standard deviation
(3) Minimum to maximum population
(4) Name of the Malay village from which I sought the camp, not of the camp
(5) The camp was adjacent to a block of an oil palm estate named Felda 6

more time is spent on such activities; rather these activities are undertaken even when there are competing income-generating opportunities. They have high cultural value. Importantly, hunting-and-gathering activities constitute a critical route through which children develop early foraging skills and come to know the forest. Seasonal activities are honey- and fruit-collecting. I will describe some of the major components of the hunting-and-gathering economy more extensively in chapter 6.

The main source of cash income is commercial extraction of forest products: primarily rattan (Batek *ʔawey*; mainly *Calamus* spp.) and eaglewood or gaharu (Batek *baŋkol*; *Aquilaria* spp.). When opportunities arise, men may do some day laboring and occasionally there is some casual planting of fast-growing vegetables. Full-blown agriculture (permanent field farming) is the least favored of activities; the Batek dislike the monotony of the work. Those living close to the headquarters of Taman Negara national park are also involved in tourism, both in hosting the visits of tour groups to their camps and settlement (Was Yɔŋ) and in guiding tourists to the summit of Gunung Tahan. The former is an image-selling enterprise that involves whoever is present when tourists come; this is supplemented by the manufacture and sale of bamboo wares (blowpipes, quivers, darts, combs, etc.). The latter attracts primarily young and middle-aged men, who may come in from other parts of Pahang to take advantage of opportunities. In recent years, one small dimension of tourism is when individuals or groups are taken out of the forest to put on exhibitions and performances, for example to promote a hotel. So far this seems marginal.

Kirk Endicott has argued that there is an economic logic to the mix of activities: those the Batek prefer give the highest returns for labor expended. Moreover, there are social pressures to contribute and share, which I call a background chorus of expectations. These do affect procurement behavior. Someone who does not like to go to the forest would be considered lazy (as marked by one word, *tasoŋ*, whose sole meaning is "lazy to go to the forest"). In the negative assertion, they say they want to go to the forest because "I don't feel like sitting in camp." In the positive sense, they express an intention to go out and do something, whether to look for food generally, or to hunt, dig, fish, etc. This is how they talk about work. Only one activity is explicitly identified as "work:" commercial collection of rattan or *kərjaʔ ʔawey* ("working rattan," from the Malay *kerja* [work]).[19]

Egalitarian norms of behavior characterize social life. Leadership is situational and activity-oriented rather than ascribed. Decisions are made at the level of individuals; group decisions tend to be the best possible (though not infallible) compromise between blissful consensus and anarchic dissension. "Bossing" occurs but is frowned upon; even children have the power to object and refuse compliance. Following a typical hunter-gatherer pattern, conflict and tension can be resolved through physical withdrawal; for example by moving away from disputants. There are some beliefs and practices that protect individual autonomy. Forcing another person to do something he does not want to could cause that person to fall sick.

The primary unit of production and consumption is the nuclear household. Interhousehold sharing of food is the primary idiom of social life and an intricate display of social performance. The larger wild game, the most "public" of all procured foods, is shared among the members of the hunter's household, their closest kin (usually the primary kin of the parents), and, if portions remain, to other members of camp. Wild yams, fish, and forest fruits are less shared, though are never withheld from anyone who asks for them. It is better to give than to ask, and social life is a dance between making things available to others and retaining control of the products of one's labor.

The government agency tasked with administering the Orang Asli is the federal-level Department of Orang Asli Affairs (in Malay, Jabatan Hal-Ehwal Orang Asli, the JHEOA); its raison d'être is the assimilation of the Orang Asli into Malaysian society. Assimilation is expressed as assimilation into Malay society through Islamic conversion. Probably the majority of the Kelantan and Terengganu Batek has converted to Islam (*maso? gob* [lit. entering Malayness], from the Malay *masok Melayu*); we don't know to what extent this is just "nominal." JHEOA was established in the early 1950s as a response to the exigencies of the Emergency (counter-insurgency war; 1948–1960). Prior to that, the colonial authorities had taken a "benign neglect" approach to aboriginal administration. The Emergency was largely fought out in the forest, with a number of aboriginal groups providing food and aid to the communist terrorists. Authorities recognized that winning the "hearts and minds" of the aborigines was critical to military success. Hence the establishment of the Aborigines Department (the forerunner of JHEOA) and the introduction of the official category Orang Asli to designate all indigenous ethnic minorities in the Peninsula. The Batek first came under the department's attention around 1956 or so, when an attempt was made to "regroup" Kelantan groups and persuade them to settle down and take up agriculture. The experiment was not successful.[20] The Batek, however, retain fond memories of the British period, for they were given food rations and medical aid. Today, one practical problem for JHEOA officials, in the face of the Batek's extensive land use system, is that the people are not all concentrated in one place all the time and cannot easily be reached. This makes the provision of government services a challenge that, however, can be surmounted. It does put the people out of the direct control of the state.

Following the Path

To return to the opening narrative, I was delighted with how this fieldwork was launched. Fieldwork, as I was to learn the hard way, does not usually have so much shape. It's one long process of uncovering information, searching for patterns and connections, and collecting, sometimes at dull, decelerated, monotonous speed, bits and pieces of answers to the questions we ask. Shape comes only later, after fieldwork, when it's time to organize information coherently, make insights concrete, analyze, and write.

In 1993, I was fresh from my first year of graduate studies at the University of Hawai'i in Honolulu. Everyone I knew at the East-West Center, which funded my studies at the university and where I lived in dorm with other student Fellows, was still doing fieldwork, just back from the field, or preparing for the field. No matter what their disciplinary focus, be it in the humanities, social or biological sciences, they all seemed to take it for granted that fieldwork was *the* incomparable medium for gathering data and contributing to knowledge. To me, this was heady stuff. I badly itched to have my first try.

So I read up on the Orang Asli and returned to Malaysia to look for "my people." I hoped to spend the summer with a little-known people, the Batek Nòng (see "Notes on Names"). The project I had designed involved studying environmental perceptions through the analysis of folklore. A series of fortuitous misdirections intervened, and I found myself trudging up a logging road and walking into a Batek camp, my first ever. I never did find the Batek Nòng but I did not need to. I knew I had found my fieldsite when I saw that camp and that logging road, side-by-side, defiantly challenging me to rethink everything I thought I knew about tradition and modernity.

Now, with the intervention of ʔeyTow et al., the Batek had given me a direction; all I had to do, it seemed, was follow the path. To where it would lead me, I had no idea at the time; the nuances and implications were not clear yet. At once, though, I was intrigued. That idea of the "*surat*"—the letter—for one thing. The Batek had a story to tell, me to tell it to, and they were encouraging me to see the equivalences between their oral literature and the written word. And to extend the reach of the *surat*, I was given a practical role: a courier of information who crosses boundaries between forest and town.

Oral literature does have a letter-like quality, and in the way it works it bridges both time and space. "Passed on down, all the old old people, passed on down from one to another. Down down down down until now, until when, until what year we have this world. All this happened in this world. If that was a different world, maybe I couldn't tell this story," as ʔeyJudiy, a man about twenty-four years old, told me a few weeks later when he narrated the myth of the world's origin.

To paraphrase, the stories are passed on from generation to generation, from before till now, from now till forever, till the end of the world. If the events he described had happened in a different world, the stories could not exist. That this world exists gives the stories their continuity, their meaning, and their validity. By assigning the role of courier to me, the Batek imply that I and, by extension, my audience, have a place in their world. Because everyone belongs to this world, which to the Batek is centered on the forest, so must everyone hear and recognize Batek stories.

Lest one imagines that my visit to the Batek had in some momentous way instigated them to redefine their notion of the world, here's ʔeyGk two years later on December 23, 1995, months into my second research period when I was doing work for my dissertation. I had moved beyond stories. One of the side paths I took was to examine the roles of shamans; shamans are all-important in the organization (and to some degree systematization) of cultural knowledge. I was interested in shamans less for their knowledge than for their roles as leaders. Kirk Endicott had already written a monograph on the religion, which includes much discussion about shamanistic knowledge.[21] Christian Vogt, who worked with the Batek from 1994 to 1995 (dissertation still in preparation), had taken off where Endicott's study ended and also focused on the religion. I saw no need to duplicate these excellent studies.

The shaman, ʔeyGk said, looks at the "eye" (*mẽt*) of the cosmos. His (or her) main concern is the safety (and I take it the sense of security as well) of the people; safety from threatening skies, ground subsidence and landslides, attacks from predators like tigers and elephants. In his dreams and trances, he looks for the right way to do things, looking everywhere, in the world beyond the forest, the world of the Chinese, the white people, the Japanese.

As ʔeyGk elaborated, the shaman does not and cannot know everything. But his *vision* has to be all-encompassing. The purview of the shaman includes the world outside the forest. And the shaman guards the world from harm because he is the first to find out things and he is perfectly poised to warn people when destruction is afoot. The degree to which he succeeds in guarding the world is the measure of his prestige and the source of people's faith in the ways of the forest. People can be disobedient and heedless. Part of the shaman's role is to tell people what they are doing wrong and what will happen if they transgress the moral order that deities like Gubar represent.

I mention all this because the Batek's desire to reach out to the urban world might come as a surprise. It's easy to overlook the fact that the Batek vision of the world is quite extensive and inclusive, rather than introspective. The Batek's pattern of tracing affinities between the forest and the broader world is delineated in a central myth of origin, known to them as the *bakar lalaŋ* (burning of the grasslands) story:

> In the beginning, all *baŋsa* ʔ [races, ethnic groups] were the same. All were Batek. One day, Tohan's younger brother ʔadam set fire to the *lalaŋ* [grass, weeds] where the Batek were living. . . . One family fled into the forest. They left behind their *surat kitab* [religious books or papers]. They got burnt and that's why their skins are black and their hair is crinkly. . . . Another family was near the riverbank. They jumped into the river and took the *surat kitab* with them downriver towards the sea. These became the Malays. The Batek looked for their *surat kitab* but the Malays had hidden them. This is why the Malays pray now.[22] So the *surat kitab* belonged to the Batek first.

Versions or variants of this myth or its structure can be found among many indigenous peoples in Southeast Asia.[23] More pertinent for the present, the myth explains how an original population of one became differentiated into the ethnic mix of today and shows how the Malays are the Batek's "most important outside reference group."[24] The Malays (called *gob*) are the politically dominant ethnic majority in Malaysia and are never far from the Batek's thoughts; almost all the government officials and local villagers they encounter on a regular basis are Malay.

In other tellings, the Batek will tack on the names of other groups whom they have met over the years. Though the Malays are the prototypical outsiders, the Batek say that the Chinese, white people, Japanese, and so on also originated from this fire. What is interesting about the myth, which resonates with the theme of this book, is the idea that *surat*—that key concept again—standing for writing, knowledge, and wealth, *also* came from the Batek.

In short, whether we look to individual opinion, cultural notions, or mythic markers, to the Batek the world beyond the forest is and always has been part of their history. In older days, that outside world would have been represented by Malay villages on the forest margins; now, with more people like myself visiting them (researchers of various kinds and stripes, government officials, tourists both local and foreign, etc.) and they themselves becoming more urban-savvy, the towns and cities have been brought into the Batek definition of the world. Many individuals have never stepped out of the forest but the forest is not isolated and representatives of the urban world are continually dropping in. Furthermore, the urban world is a lot closer to the Batek's everyday experience than it ever was before. I once took just four hours to reach, by truck and taxi, the capital city Kuala Lumpur from the edges of Taman Negara. In short, the Batek have an endless fund of experiences, impressions, and observations to draw from, as they examine the outside, whether for inspiration or for cautionary lessons, comparing and contrasting it with their own world, measuring themselves against it. In some sense they consider themselves to look after it. They have a certain view of their place in this world.

They identify themselves on the one hand as people whose stories are not heard, not believed, not taken up (recall the words of ʔeyTow and ʔeyBajaw on p. 2). People without influence; people on the edges; people left behind.

By contrast, the Batek never tire of stressing either how we from the world outside the forest do not have access to their knowledge. How they know the stories of the forest and we don't. How the stories of the forest are theirs and not ours. So there is a boundary of some sort between the forest and its exterior, and the Batek do see that they have something of value. They are not abjectly powerless. What they want, as we will see shortly, is to share some of their knowledge with the broader world, a process that I, as the courier and cultural broker, could participate in.

The aim of this book is, accordingly, to take up the role that the Batek suggested for me, and communicate some of their ideas. In my focus on forest degradation, the main inspiration is the shaman Tebu, who is anxious that we hear what they, the Batek, have to say about environmental problems. To Tebu's thoughts and what he means for us to know, then, we turn next.

Notes

1. The language of fieldwork throughout this first visit was Malay, the lingua franca in Malaysia which the Batek are fluent in. On subsequent field trips, I changed my research

policy and communicated exclusively—to the best of my ability—in the Batek language. As such, quotations from 1993 are translations of Malay.

2. Benjamin 1997.

3. Ooi 1963, 357.

4. Schebesta 1926 on Orang Asli classification; Benjamin 2002, 28–29 on Aslian history.

5. The population of Malaysian citizens in the 2000 Census was 21.89 million. The Orang Asli population, according to 1997 government figures, was a disputed 106,131 (Nicholas 2000, 13–14 n. 1).

6. Ruslan 1990/1991.

7. Gomez and Jomo 1997, 1. See Cooke 1999, who examines the context with Pahang timber licensing in pp. 121–23.

8. Ministry of Finance reports, cited in Teh et al. 2001, 233.

9. Sahabat Alam Malaysia 2001, 10.

10. Teh et al. 2001, 240.

11. Sahabat Alam Malaysia 2001, 19, 23.

12. Abdul Rahim 2001, 307–8.

13. Teh et al. 2001, 223.

14. Cited in Sahabat Alam Malaysia 2001, 18–24.

15. Sahabat Alam Malaysia 2001, 123.

16. Teh et al. 2001, 223.

17. Sahabat Alam Malaysia 2001, 19.

18. Kirk Endicott 1995, 245.

19. Karen Endicott 1979, 62.

20. Recounted in Carey 1976; Carey was Commissioner for the Aborigines at the time.

21. Kirk Endicott 1979a.

22. A reference to Islam: Malays are largely Muslim, and strongly identified by themselves and others as such.

23. See, for example, Borie 1887, 291; Father Dunn 1992, 29; Evans 1923, 146; 1937, 161–62; Keyes 1979; Nishimoto 1998; Schebesta 1973, 89, 216–17; Sellato 1993; Skeat and Blagden 1906a, 346–57, 536; 1906b, 219, 378.

24. Kirk Endicott 1979a, 86–87.

2

Communicating Degradation

What does Tebu want us to know? He is worried about the long-term effects of forest clearance. For someone who has lived all his life in the forest, watching his family grow up, the quality of the forests is obviously an issue of intense personal interest. And to the other Batek individuals, who share Tebu's sorrow and bitterness with the evidence of degradation around them. However, that is not the end of it. Tebu wants people outside the forest to recognize the far-reaching effects of forest degradation. The Batek acknowledge that environmental conditions—particularly conditions of land and water—have deteriorated. If this trajectory persists unabated, Tebu asserts, humanity can destroy the world. A challenging and (to some) perhaps renegade idea is that problems in Batek forest can be linked to the future of the world, even those parts that are not obviously linked ecologically, geographically, politically, and economically to Batek forest. What that "world" might be and who in it should listen to the Batek are some of the underlying questions in this book. Tebu has a practical solution in mind. He urges us, people outside the forest, to think about the consequences of taking too much from the environment and consult with them, the people in the forest, in planning for the future. The text that I recorded from Tebu, in sum, complexly connects "local" with "global" and environmental discussion with political rhetoric, all interwoven with his perceptions of the knowledge, roles, positions, and aspirations of the Batek.

It is no longer uncommon for indigenous leaders and shamans to address the world, whether face-to-face in carefully choreographed "events" and political platforms, with cameras clicking away hungrily, or indirectly through third-party channels like myself. The political value of direct encounter is of course widely exploited. During my fieldwork, the Interpretation Centre of Taman Negara contained a photograph showing the then prime minister, Dr. Mahathir Mohamed, "meeting" the people, the people in that case being not the Batek but ceremonially garbed indigenes from Sarawak (in East Malaysia). The message is delineated by the fact that Taman Negara for a long time was *the* national park in Malaysia (as shown in the name: *taman* means "park" and *negara* means "nation") and still carries strong symbolic importance for many Malaysians. The photograph seems to convey a strategic message, that this smiling leader is concerned for the people and the people have a place in the state. Such images, in other words, foster the idea of the benign and beneficent state, and one whose reach extends to remote corners of the country. Whether the promise of such encounters is realized in practical action is, however, a different matter. Tebu's decision to speak to the world, not through Mahathir or his representatives but through me, betrays how he views the Batek's position in the state. It is also telling that the photograph displayed is not of the prime minister with the *Batek*, which would be logical considering that they are the aboriginal people of the national park, for the reason that there is no such photograph; Mahathir never officially met the Batek.

Nicholas describes an interesting response to official neglect:

> On 22 June 1999, during the Prime Minister's visit to Bukit Lanjan [Selangor; home of the Temuan]—partly to campaign for the general election to the Orang Asli village-heads assembled there—several Orang Asli took the opportunity to thrust memoranda and various letters (of protest, application and appeal) into the Prime Minister's hands as he walked the red carpet to the dais.
>
> The Prime Minister was visibly displeased with having to shake hands with Orang Asli leaders and having envelopes stuffed into his hand at the same time. Standing behind me, one representative of the private developer organising the event was overheard to have commented that the Orang Asli were being very rude and disrespectful to the Prime Minister. However, to many who were present, it clearly reflected that the Orang Asli had much to voice to the national leader and that existing channels for doing so were not there or that they did not have any impact.[1]

While to the prime minister the occasion served a pragmatic purpose, to woo voters, to the Orang Asli the occasion meant, perhaps, something else. What the Orang Asli did was to *invert* the role expected of them: namely, to pay due respect to the leader, to listen, to learn, to accept, and not to protest. Before the leader could sit on a dais and dispense political wisdom from on high, he would have to run the gauntlet of those letters, reminding him that the road to development success is paved with the plight of people "below." Symbolically, the Orang Asli made the prime minister one of their own, reminding him that if their votes count, if they belong to the state, then their concerns are the state's too, and the letters pointedly show that the state has not addressed those concerns. The Orang Asli appropriated the symbols of external power: they made their problems intelligible in a form and context that the prime minister could not ignore—publicly, in writing, in his language, and addressed personally to him. The message is that, whether the prime minister liked it or not, he should attend to the people's appeals for help, even if this meant sacrificing the niceties of protocol.

The specific performances, intentions, and effects of indigenous role-reversals and inversions of dominant symbols of authority and power differ from context to context. Nevertheless, there does appear to be a growing number of forest peoples around the world, independent of any outside intervention, demanding wider recognition of problems besetting their local environments and trying whatever means they can to make their voices heard. A statement for anthropologist Roy Ellen from a Nuaulu acquaintance in Seram, Eastern Indonesia, provides an example. The statement (dated 1994) was recorded by a colleague of Ellen's, doing fieldwork in the area, to be handed to Ellen, back in the United Kingdom; the concluding passage reads: "Therefore if Roy can find a little help and wants to talk to the officials here in Indonesia I ask that he help us a little so that they do not come here and work again. We do not want them to because we are already suffering a lot."[2] As Ellen interprets the text as a whole, which is a critique of the effects of logging and land loss, "From a position in which Nuaulu saw themselves negotiating *with* the Indonesian state, they are now simply citizens *of* that same state. There is an acceptance that events are no longer under their own control, that they can no longer take them or leave them."[3] What Ellen does not discuss is why the author considered that the foreign anthropologist is in a better position to stop officials or influence events. After all, it could be argued that as citizens, the Nuaulu have equal entitlement to the channels provided by the state and need not go beyond those channels to foreign advocates.

There is another example rather closer to home. Over in Sarawak, Penan blockades against logging companies in the 1980s were celebrated in media headlines around the world. Penan became "an icon of resistance for environmentalists worldwide."[4] Popular discussions of the Penan blockades inevitably link them to the name of Bruno Manser, a Swiss who lived among them for years, learnt their language, was instrumental in organizing the Penan to resist logging on their lands, and was most responsible for popularizing their cause abroad. What links the Penan story with the Nuaulu and the Batek, however, is not the presence of sympathetic outsiders so much as how the peoples see their positions relative to the state.

From the point of view of government officials, there are official channels of communication that anyone can take advantage of. Indigenes can even, under controlled circumstances, address the nation's leader. From the point of view of the people, however, these channels either do not work or do not serve their purposes and so they must develop alternative means of advancing their grievances. Brosius has recorded how the Eastern Penan describe the customary remoteness of government representatives:

In describing why they have erected blockades, Eastern Penan provide one reason more than any other: They feel that their voices are not being heard. Among the most common refrains among contemporary Eastern Penan is that "the government does not hear what we say." Their expressions of frustration about this convey utter exasperation. One man stated that even if his mouth ripped open all the way down to his feet, [logging] company people would still not hear what he said to them. Another said that talking to representatives of timber companies or the government was like talking to a drawing: They neither hear nor respond.[5]

The Batek have not expressed their frustration in the same way. It's inevitable that they'll be compared to the Penan: both are marginalized forest-dwelling hunter-gatherers whose lands are coveted and appropriated by powerful enterprises. Ahead of that, I should point out how they are different.

According to Brosius, the Penan hardly see any difference between logging operators and the government: "They hold two parties responsible for the present situation: logging companies and the government. In fact, they often conflate these, assuming either that they are a single entity or that logging companies are working under direct instruction from the government."[6] The Batek, by contrast, seem more exposed to subtleties in the government machinery, probably because government administration in the Peninsula is centralized under a single agency, the JHEOA, whose presence is well-established among Orang Asli (see p. 14). There are also the wildlife rangers and officials that run Taman Negara, and the occasional officials from the medical services and state government.

More important, the Batek have not organized acts of protest against logging. Logging, to them, does not seem to have the symbolic weight that it does for the Penan. As we'll see in Tebu's message, it is not logging per se that alarms them so much as the full-scale transformation of the landscape. Logging is a presence in their home that (like Nuaulu) they know they cannot control and they have got used to. They do share the Penan's frustration to see the land churned up and important markers of home and identity, like trails, campsites, burial sites, and significant plants, mowed under, and we will examine these feelings later (especially chapter 5). To a degree, such changes can be tolerated. What really worries is the spectre of total destruction.

Two principal symbols of destruction, I venture, are monocrop plantation development and hydroelectric dam-building. For these changes are irreversible. As described in chapter 1 (p. 8–9), plantation development replaces the original floristic diversity with monotonous rows of (usually) alien tree species. There is nothing culturally and ontologically familiar in a plantation forest. Dam-building demands sizable catchment areas and floods vast hectares of forest. To a people whose home never contained any body of water that was so extensive that one could not see from one bank to another, the shock of seeing forest flooding at such a scale is tremendous. There is nothing in history to compare it with.

Tebu's message, then, has to be understood in relation to two plain facts. One: his home, the forests of Pahang, is threatened; he fears that the degradation may never end. Later in this chapter I will examine the text of his message. We will find that he draws on traditional philosophy to relate the causes and implications of degradation to a rich set of ideas concerning the proper place of people in the forest, and the place of forest in the world. Two: like Nuaulu and Penan, and probably countless other forest communities in Malaysia and around the world, Tebu recognizes a failure of communication between his people and the state, and by entrusting me with his thoughts, he hopes to reach those who can prevent further forest loss.

Unlike the Penan, the Batek are very much "off the beaten track." They have not been visited by environmentalists seeking to use their plight for public relations capital. They *have* been visited by Euro-American tourists who found themselves impressed with the Batek's "noble savage" way of life. Many Batek dismissed such romantic notions to me. Throughout my fieldwork, whenever I struggled to master even the simplest of tasks, people did not fail to remind me how difficult life is for them: compared to the push-a-button-and-make-things-happen culture of urban Malaysia, that is. If they are aware of the global reach of environmentalism or the kinds of political linkages that environmentalists build with indigenous communities[7], it is an awareness of the haziest kind. In short, Tebu is not strategizing a way to jump on the indigenous people's bandwagon.

In general, the Batek are relatively uninterested in political movements and activism. As we'll find in chapter 5, their history has been one of resisting entanglement or entrapment in other people's schemes and projects. The sole political organization of Orang Asli, the Peninsular Malaysia Orang Asli Association (in Malay, Persatuan Orang Asli Semenanjung Malaysia; POASM), counts no Batek among its membership. When Batek individuals travel outside the forest, sometimes to give "cultural performances" for the tourist trade, they do not come back fueled with ideas of how the outside world is waiting to embrace their political participation. Rather, their impressions are cultural and ethnological. On present evidence, I think it unlikely that environmentalism among the Batek has exogenous roots, whether those roots be from domestic or foreign conservationists and environmentalists.

This point, the independence of Batek thought and perception, is necessary to stress. As we'll find in chapter 5, there is in Malaysia a long-held patronizing attitude towards indigenes like Batek and Penan; indeed, of anyone who seems "primitive" and "traditional." Two common terms of abuse that the Batek bristle at are the Cantonese' (Chinese) *samfan* (forest barbarians) and the Malays' *sakai* (slave; dependent; refers only to Orang Asli). Both of these words are still commonly heard. As Hood warned, "We must be aware and sensitive to paternalistic orders which try to frame the problems of tribal peoples as if they are in need of civilizing."[8] This civilizing imperative, he writes, is "mere camouflage" that distracts attention from the problems that need to be solved, structural inequality being one. One aspect of this approach is to trivialize whatever the people say or to express disbelief that "primitives" could ever come up with sophisticated thoughts and actions on their own.

As Jomo reports, when the Penan blockades first received broad attention, "Malaysian politicians and the international media—anxious for a modern Tarzan story and contemptuous of native abilities—attributed [Sarawak natives'] increased resistance to Manser, ignoring the sustained build-up of such opposition over the years, especially from the early eighties, i.e. before Manser's appearance."[9] The Batek have already experienced some of this. Just before my dissertation fieldwork, the newspapers had a field day accusing (without basis) Christian Vogt, a German anthropology student, of "instigating" the Batek to demand their rights from the Taman Negara Resort and tour operators that had been exploiting them.[10] Both the media and investigating officials from various government agencies, rather than focus on the Batek's economic grievances, were most concerned with whether and how the white man had "instigated" them—a clear example of the kind of "camouflage" exercise Hood critiqued. In the present case, all I can say is that this book would have taken a different shape had Tebu not instigated me with his brand of environmental insight, political urgency, and moral imperative.

The authority of this account depends on convincing readers, especially Malaysian readers, of the context in which my relationship with Tebu and the rest of the Batek developed. The skeptic might even suspect (so I have been warned) that Tebu is a fictional character, whom I have invented for the purpose of voicing my own vision of the environmentalist agenda. It is important for that reason to present Tebu as I know him and as much as possible in his own words, so that he comes through these pages with his humanity, relationships, and character intact.

Let us begin with the evening of November 4, 1996, when Tebu communicated his message to me. Two days short of my departure from the forest; my dissertation fieldwork, lasting some fifteen months, was over. We were in Was Yɔŋ in Taman Negara. Tebu was the only man present who was also in that original storytelling group of 1993 (see chapter 1). His group (five men, six women, eleven children altogether) had joined the Was Yɔŋ community two days earlier.

I last saw them at Ralat, a tributary of the Kɔciw, several hours' trek from our 1993 campsite (see map 7). We parted ways in early July and had many hopes of crossing paths again. The fruit season was arriving and with this change of tempo, our camp group was breaking up and dispersing to different locations around the forest. Tebu's group moved on to game-rich portions of the Tɔmaw (Malay Temoh) river valley, midway between the Kɔciw and Tɔmiliŋ river systems. I, on the other hand, would travel along the Ruwiw (Malay Ruil) river system, southwest of Was Yɔŋ, with another group, the "Ruwiw group" (see map 8); we wound up the season at Yɔŋ. In these final weeks, there was much to-ing and fro-ing, and lots of old friends turning up unexpectedly. Tebu's group had sent word ahead, finally arriving on November 2. That day, the population surged to a high of seventy-two.

So it was with expectation of catching up on news that I visited Tebu's family on November 4. They were bunking casually in one of the vacant houses as they did not intend to stay long. It was about 10:30 PM. Almost everybody was still awake—chatting, smoking, drinking tea, bedding down on sleeping mats. An ordinary scene of nighttime conviviality. A glowing hearth was off to one corner of the room, and some kerosene lamps flickered around the floor. Tebu, I had come to appreciate after the Ralat travels, is much respected for word and deed; he is an exemplary character. Rare for me was the opportunity to see him, let alone to talk for any length of time. There were a few brief meetings but I was able to live with him and his group just twice during that dissertation year (July 22 to August 19, 1995; March 29 to July 5, 1996). Both times, our camp groups were large: the average population was over sixty (see Table 1.2). Tebu was always busy with his own activities, like preparing for the fruit-season rituals, and I found it difficult to talk quietly with him. Any time we got a chance to talk, someone else would butt in, and the moment would be lost. Moreover, though friendly enough, Tebu was not the most forthcoming of persons to me.

For a long time, I did not know him as a *hala?* (shaman). He fits the picture conveyed by Endicott's description of shamans:

> The role of shaman is just one of many social roles played by the average mature Batek; it does not set one off as a special type of person. Even among the Aring Batek [of Kelantan], who sharply distinguish shamans from ordinary people, the shamans do not stand out from the crowd. It has often been reported that Negrito shamans do not reveal themselves to outsiders, and the identities of the shamans can only be found out indirectly or after long acquaintance with a group. . . . [Shamans] are not socially or psychologically aberrant in any way.[11]

In my memories of Tebu, I see him as a husband grumbling, "I wanted to go to the forest today but my wife wanted me to move the lean-to" (he did move the lean-to, after much glowering contemplation over the task—this is the Batek equivalent of a wife ordering the husband to do a spot of house-cleaning or renovation when he would rather be elsewhere) or telling his wife not to fuss over the younger folk's stereo-playing; as a surrogate father flashing a concerned look at an orphaned ward; as an irritated grandfather offering his arms to a wailing grandchild; as a group leader organizing the route in a camp move. There are also brief glimpses; like the time I and two young men had lost our way in the forest when he came purposefully down the path, and scornfully pointed us in the right direction.

It was not until the fruit season of 1995 when Tebu organized some relevant rituals that I could discern his shamanistic persona. *Discern*, but not confident enough to class him as a *hala?*; even the Batek themselves cannot always say. A shaman cannot run around showing off or boasting about his knowledge; nor can others freely identify his shamanistic status and discourse on his powers. Such open displays invite punishment or lead to a loss of efficacy. One of Tebu's fellow shamans was wont to say things like: "I don't know anything." "I only know a little bit." "I hear the old people say it could be that way." All discussions about living shamans must be conducted using oblique references of this sort. It was a long time before I was sufficiently adept at recognizing cues and interpreting hints. Until then, seldom could I tell who was and was not a shaman.

The anthropologist in me was mindful of last-minute interview questions that evening of November 4. We began with small talk and cheerful teasing. I showed off scraps of new information in my repertoire; I asked to hear some myths again. Myths of crops, plants, animals, the snake, and tiger—stories long familiar to me but that always needed one more telling. Typically, whenever plants and animals enter a conversation, the topic of taboos quickly follows. We chatted briefly about these. The list is endless and I was not eager to launch a fresh assault that night. The noise inside the house was tapering off.

Eventually, only Tebu and son-in-law ʔeyKaw kept the conversation going. Tebu had the most to say. The dimness of the room was a death knell to me; I desperately wanted to sleep too. But Tebu, who is in his late forties or early fifties, had impressive stamina and was wide-awake, sitting up. Young ʔeyKaw, lively as ever, lay on his front. I was loath to shorten our chat or even to disrupt it by sliding into a horizontal position. They had questions they wanted to ask me, ideas they wanted me to remember. I did not use a tape recorder. I jotted notes in my scratchpad as Tebu spoke. Sometimes I interjected with brief questions and comments. In this wearied state I tried just to write down as much as I could, and much of that in his language and idiom. The resulting transcript (appendix A) is a multilingual hodgepodge.

It was almost three in the morning when we stopped talking. By then we had ranged over many topics and issues. In what follows, I present what Tebu said he had been waiting to tell me. Following an extensive reanalysis of my notes, this version of the message differs from (and supersedes) earlier ones.[12] Readers wishing to examine Tebu's language should consult appendix A, which is a transcription of his message. To convey something of the topography of his mind that night, and preserve the movement of his thoughts, I have not rearranged the chronological order of the extracts that follow; these are interleaved by my own commentaries. I hope also to introduce some of his and the Batek's core concerns, which frame the selection and organization of this book's material, and to provoke questions in the reader's mind. More extensive exposition continues in the chapters that follow.

Tebu, like the Kəciw group back in 1993, hopes to speak through me to the people outside the forest who can be persuaded to heed and value his argument. (As he was in the 1993 group, I cannot avoid the suspicion that he had engineered some of what they told me to do.) He never spelt out who in this world, or what social and political sectors, he intended his message to reach; perhaps he did not have a concrete idea, since he has never been to the cities and, needless to say, has never been invited inside the halls of power. The ideal audience, then, is but a concept, vaguely defined. That is the role of this book; to attract as many people as possible to his thoughts, ideas, and warning.

In the following discussion, the pronoun "we" (except when specified otherwise) refers to the world that I'm part of, beginning with the readers of this book. My role is to interpret: I draw on my prior knowledge of Batek idioms, stories, behavior, experiences, and concerns to help render Tebu's thoughts more accessible to readers. Sometimes this involves straightforward restatement of his ideas in my own words. At other times, I provide background context to relate his

Plate 2. The Batek are alarmed by the full-scale transformation of the forest landscape, as revealed by new roads like this one, which fringes the southeastern boundary of Taman Negara National Park. July 1996.

observations to what is going on in the Malaysian environment today. At the most metaphorical moments, I offer analyses of his idioms and images. The route to clarification may seem digressive at times; each of my commentaries should be thought of as a topical section that stands on its own yet adds to the overall meaning of the message. Readers impatient to read Tebu's message in full should therefore turn first to appendix A.

Tebu's Message

For Tebu, the most critical issue is the future:

> *Take away the forest, the* dəɲaʔ *[world] ends. We want people to know that the world can end. Already there aren't trees.*

His tone is insistent. *We want people to know. . . .* He links the fate of the forest to the fate of the world: *the world can end.* To save the forests is to save the world and, as he'll make clear later, one way to start is by listening to what they, who have observed, studied, analyzed changes for a long time, have to say.

His example of change is vivid and hits just the right note with me:

> *When they dynamite, Gubar makes it rain a long time. We remember.*

He is referring to the building of a road on the other side of the Tembeling River (see Plate 2). The road-builders had been using explosives to widen gullies; the deafening *booms* of these explosives punctuated my final weeks in Was Yɔŋ.

The roadworks were the latest phase in a process that began in the mid-1980s. Before then, there was no direct land link to this area from the market centers and the river was the effective highway. Then an earth road, sixty-five km long, was completed between Jerantut and Kuala Tahan. Originally accessible only to heavy-duty and four-wheel drive vehicles, improved surfacing of the road had by 1996 made it possible to bypass the three-hour boat ride from Kuala Tembeling and drive directly into Kuala Tahan. Eventually this road will replace the old logging tracks.

To be honest, the road-builders were not the first to alter that landscape and open up forest patches. Long-established Malay villages like Kuala Tahan and Kampong Tekah fringe the riverside and for these interior communities road links are a blessing. Not only can they aim for *in-situ* development, they can reach medical care and other social services much faster. The relatively few remaining *neram* trees (Batek *kəmajay, Dipterocarpus oblongifolia*) overhanging the east banks of the river are physical reminders of what the environment used to look like. Road-building is perhaps change of a different order of magnitude, though. For ecologists and for the Batek alike, forest clearance exposes the ground; erosion and river siltation follow. But, at this moment, it is the auditory environment that worries Tebu—the blasts of dynamite. Listening to him, I was reminded that loud noises could bring ecological harm as well. These noises annoy Gubar, the thunder-god from whom emanates noises of the sky: *Gubar makes it rain a long time*. It's like Gubar is not used to having *his* domain—noise-making—challenged by the activities of people. Loud noises have the power to disrupt the equanimity of the other-worldly realm. In his anger Gubar sends down more rain. In short, Tebu linked the *booms* of explosives to rainfall patterns. Every *boom* invites an angry response from Gubar; the thunder comes, rainfalls last longer.

And, he remarks, they have observed the patterns of change, and they "remember."

After thinking a moment, he turns to me and continues eagerly:

> *Don't take away more forest, make that the limit. We* [inclusive] *look for food alike. We* [exclusive] *want to* pakat [Malay: decide what to do together]. *We cannot be overly covetous—we should take only what's appropriate for our livelihood.*

Now Tebu is talking directly to his audience. He has thought about and made some linkages, relating cause to effect. Environmentalists will appreciate the message: Tebu wants us to recognize limits, to *take only what's appropriate*.

The subtlety of this passage is lost in translation. The English language does not distinguish between the inclusive and exclusive forms of "we" (from hereon, these distinctions are so marked in the transcription). Batek (like Malay) does:

Tebu has switched pronouns from *ʔipah* (exclusive: we but not you) to *heh* (inclusive: you and us together). In using *ʔipah* Tebu speaks up for the Batek's side of things. In using *heh*, he identifies common cause with his audience; he brings the rest of the world into the Batek's side and therefore urges us to go beyond surface differences and accept what's common to them and us. The most fundamental similarity is the need for sustenance: *we look for food alike*. With this as the basic principle, everyone is enjoined to observe the same social limits: *We cannot be overly covetous—we should take only what's appropriate for our livelihood.*

He is perhaps judging the behavior of outsiders by the moral standards of the forest. The capital irony is clear to him: the search for food may be universal but access to food is not. The more *we* (outside the forest) prey on natural resources, the less there is for *them*. It is they, the forest people, who must bear the brunt of the effects. When outsiders look for food, the forests are liquidated for capital—wood products, minerals, oil palm estates—that are converted into even greater wealth fueling even more capacity to take food from the forest.

Now at this point it is well to consider that Tebu is partially metaphorical. As the transcript (appendix A) shows, he has eschewed the Batek phrase *cam bab* (to look for food) for its Malay equivalent *cari makan* (lit. searching to eat). He stresses the food metaphor with the Batek verb *radiʔ*, which means to be overly fond of something, principally food (which I translate here awkwardly as "overly covetous"). Why the Malay substitution? *Cam bab* is only about the search for food: foraging, the food quest, the satisfaction of nutritional and gastronomical needs. It is not a metaphor like *cari makan*, which is a common phrase among Malays to refer to the search for economic opportunities or anything that will keep the hunger away. *Cari makan*, from its original peasantry meaning of keeping hunger away, has become a justification for any activity that will give a person a leg-up in life, including corrupt and unscrupulous behavior;[13] and Tebu has merely extended its usage to include forest exploitation.

Whatever might be Tebu's *reasons* for using the Malay words, one effect is to signal exactly what he means when he condemns forest cutting: that the search for economic advance has gone out of control and it is time to put a lid on excess. For at the heart of it there are people who do literally eat off the forest. So he makes a plea for consideration from us. If they must live with the consequences of our actions, then they should have a say: "let's *pakat*." He wants us to talk with, and listen to, them. He is asking for a voice in decision making, in planning how to make a living from the forest that will not jeopardize *their* economic opportunities.

Təmaw River already has no trees. Only oil palm.

This, again, is for my benefit. It is another example of what has been happening to the landscape. Tebu's group has just come from Təmaw River, one of the long-flowing tributaries bridging Lipis and Jerantut districts. I had heard much about it. There was naʔGk reminiscing one evening about being a young wife traveling around the Təmaw area with her parents while her husband lived on his own in Yɔŋ. That would have been some time in the mid-1980s. That kind of experience always creates a certain bond; people remember these places of the past and how they are linked to important biographical events and they say they feel *ha?ip*— longing and nostalgia, a sentiment we will hear about again (chapter 5).

Before Tebu's group moved there this year, his daughter na?Kaw badly wanted me to go with them and whetted my appetite with stories of how many kinds of game and fish are found at Təmaw. So abundant that they never want to reveal its riches to outsiders. Ta?Kadɔy once said if they tell the Malays what they know about Təmaw, the Malays would fish it out with dynamite. And Tebu warns me of yet more serious threats: already the oil palm estates are moving into its upper reaches.

What are the implications? He evokes the images, one after another, so that I will not miss the significance of the problem:

> *Our* ɲawa? [souls] *live upon the trees. The forest is the* ?urɛt [veins and tendons] *of our lives.* He holds up his arm, and points to the veins and tendons therein.

These are vintage Batek idioms: the soul; the trees; the forest; the veins and tendons. Here they are paired off in relations of similarity: the souls of people // the standing trees; the forest // the veins and tendons of people's lives. The main comparison here is between the forest and the human body.

When he says that the souls live upon the trees, I do not think he is in literal vein. As we'll see in chapter 4 (pp. 86–89), there is a range of arboricentric images in Batek cosmology. However, there is no evidence from their theories of life and death that souls live in or on trees. I think Tebu means "soul" in the sense of "being"; without trees, it is not possible to be human. That, after all, was the lesson from Təmaw River, that the original forest trees are losing ground to estate crops. The planters are removing the humanity from the landscape.

There is another implication, which gives a different sense of "soul." One of the myths we revisited that night had concerned the origin of trees. At the beginning of time, when the population had grown too large, there was not enough food for all the people in the world. So half the people then transformed themselves into food-bearing trees. Given these anthropogenic origins, to cut down these trees is to rob the people of sustenance, and a big part of what has made them as they are today.

Which immediately takes us into the landscape that trees themselves are nurtured in, the forest (*həp*), the "veins and tendons" of the Batek's lives. On first encounter, this is an odd image to use. Elsewhere in the Batek canon, veins and tendons do not get good press. It was the veins and tendons of a primordial masked palm civet (*Paguma larvata*) that originated the disgusting creatures, like worms, centipedes, leeches (see pp. 83, 90). Here, stressing the point with his own veins and tendons, Tebu directs my attention to the circulatory system of the body: critical to the flow of blood, that which keeps life going.

Considered thus, "veins and tendons" looks more promising. Just as we should stanch the draining of our blood, so we should do what we can to prevent the "blood" of the forest from leaking away into infinity. Poetically, the metaphor gives the sense of something that is there, right under the skin, visible, concrete, necessary, yet slightly elusive and hard to define. Whether or not he consciously intended the effect, Tebu in using this imagery brings to *my* mind the botanical structure of the forest, its complexities and interconnectedness. As Lakoff and Johnson put it, the "essence of metaphor is understanding and experiencing one kind of thing in terms of another."[14] But, as Ingold has argued, metaphors like Tebu's may also draw attention to "relational unities"[15]: Tebu does not say the forest is *like* the veins and tendons of their lives—the forest *is* the veins and tendons of their lives. The forest keeps them alive as the veins and tendons keep the blood flowing in the human body. Both interpretations of metaphorical usages have their place here. Tebu is speaking to *us*. Were we skeptics, we could appreciate metaphor as poetics (following the Lakoff and Johnson model) and we would still, I think, get the point. Whether we think of the forest as a source of poetic imagery (as "good to think") or as the lifeworld of the Batek (as "good for life"), we can't escape the hint in Tebu's rhetorics: against a reductionist approach to the forest; for example, solely in terms of the search for useful resources or food, a theme that has already surfaced in this discussion.

There are graver considerations; the souls and lives of the forest people are not the only forces at risk:

> *This* te? [land] *is an island. How can the land hold together without trees?* With hand gestures, he shows how the island would wobble.

He reminds me that the world is a disc of land; it is an island resting on subterranean waters and that is held up by trees; an island that loses substance when trees are gone. Without trees, the land loses its support, like a house without its frame, like a body without its skeleton; it crumbles in the waters because nothing shores it up. In a common Batek phrase, *te? neŋ ?uhm tahan* (the land is unable to hold up). Tebu must discuss this because he is reaching out to the world beyond the forest and from his perspective we have forgotten some basic facts of ecology: the hydrological functions of trees, the instability and vulnerability of the land.

All this comes as no surprise. The cultural idioms may seem a little exotic but Tebu's concerns are shared widely by the environmental lobby—and government agencies—in Malaysia. Land development (including the inevitable release of sewage and animal wastes into the waters), forest clearance, and soil erosion are often identified as critical causes of river sedimentation and siltation. The practical ramifications include, to quote a Ministry of Health official,

> flooding in low-lying areas, frequent occurrences of flash floods especially in urban areas, depletion of aquatic life, problems of water supply and hindrance of economic activities such as fishing.

The consequences of these effects include fatalities during floods, disease as a result of poor health care, and contamination of water supplies. Another consequence, often overlooked, is the mental well-being of people whose lives are constantly threatened by the imminent danger of floods.[16]

But Tebu's vision is a lot darker than this scientific language allows.

In the past, we [inclusive] *lived in peace, we* [inclusive] *weren't losing the world.*

Environmental change is making us "lose the world"; the time of peace and complacence is gone.

To heighten the sense of loss, he reminds us that destruction also occurs in other realms:

The hala? ?asal [superhuman beings] *they say: "the* kəlaŋes [heart] *of the earth they made."*

These beings created the world, and Tebu suggests that they remind him of what they did for us.[17] As we'll see in chapter 4 (pp. 79–81), this disc of land that formed out of the primordial waters began life as a clump of earth that slowly expanded as the waters receded. That clump of earth, I believe, is the kəlaŋes (heart) that Tebu is referring to. He is using, again, metaphor; as with the earlier image of veins and tendons, he makes a direct equivalence between the earth and the human body, and what makes the metaphor work is the allusion to life itself. Just as there is a part of the human body that regulates its proper functioning, so there is a heart to the world, which is critical to keep. There are more complex conceptual associations. In Batek theories, the seat of emotion—that with which we remember and feel strong sentiments like *ha?ip*—is the heart. For example, when one loses the way in the forest, it is because one's heart does not remember. When one is forgetful, the heart is considered to be *jɔbec* (bad), i.e., in distress or disorder. We may, therefore, consider the heart of the world to be its emotional center.

And the superhumans look at the present day world, and what we are doing to it:

The superhumans they remember they ha?ip [feel longing].

The superhumans, I would say, remember two things about the earth. One, they used to live on earth too. Kirk Endicott writes that the superhuman beings "have existed since before the creation of the earth, and it is assumed that that they have always existed."[18] As I'll elaborate in chapter 4, in creation times superhuman beings transformed themselves into, or caused the creation of, important components of the forest like various landforms, plants, and animals. So the superhumans remember *their* past, what the newly created world was like: fertile, abundant, and healthy. The superhumans also remember that it was their work that brought the world into being, which rampant environmental change makes a mockery of. And they feel *ha?ip*, which is that inexpressible feeling of longing for something or someone that is absent.

Ha?ip, if not salved, can lead to danger. It brings on *taɲol*, the state of pining away when we cannot rise above our emotions, enjoy life; we refuse sustenance, nothing can pull us out of our malaise. If we reach this stage of *ha?ip*, we can become insensible to what is happening around us. Along the same lines, the superhuman beings can become indifferent to the concerns of this world and ignore the shamans' calls for help. In other words, *ha?ip* has the potential to plunge the sufferer into spiraling depression; the next stage is death. If the superhuman beings were to perish because they are *ha?ip*, there would be no possibility of regenerating the world. The destruction of the world would be assured.

For now, this is not yet happening, because the superhumans still feel connected to the Batek:

They feel sorry for us when they hear our songs.

They still care; they have not shut down the channels of communication:

They love us so they warn us what is happening.

It is a warning that does not, or should not, fall on deaf ears:

We [exclusive] *hear what they say, we* cɛp [hold on to] *their voices.*

Let us attend to the idioms that Tebu is using; they bespeak a central mode of relating between the Batek and the spiritual forces of the forest. On the one hand are the Batek who feel alarmed at environmental perturbation. Through the medium of ritual adepts like the shamans, they communicate their concerns to the superhuman beings. These communications most markedly take the form of songs. The superhumans, themselves grieving for the forest, hear the songs and they feel sorry for the Batek. The superhuman response is to warn the Batek; if you like, to arm them with knowledge of the probable dissolution of the world. That's their role. The role of the Batek on the one hand is to continue singing, for so long as they maintain communication, they can persuade the superhumans to pay attention. More broadly, their role is to remember what they learn from the superhumans ("hold on to their voices") and to act accordingly.

We might want Tebu to spell out what exactly he is learning from the superhumans. This is not possible. As I noted with regard to Tebu's public persona, the details of knowledge obtained from the superhumans are generally kept secret. The substance of the environmental warnings belongs in this category. Clearly, as far as Tebu is concerned, such details are irrelevant. What is important is the synthesis communicated by shamans like him. We saw earlier the implicit warning against depauperated depictions of the forest. Here, in bringing on the question of superhuman beings, emotions, and environmental memories, Tebu shows us, perhaps for the first time in this discussion, some of the true complexities in the value of the forest.

Perhaps agnostics can appreciate only dimly the significance of the connections we have just seen. Tebu returns to something more mundane; he reminds me of a simple fact:

When the trees are gone, no place for us [inclusive] *to shelter.*

The forests are important for shade. This point will be appreciated by anyone who has ever sweltered under the afternoon heat or invested in an air-conditioning unit in Malaysia and Tebu means for us to do so: he is using the inclusive *heh* rather than the exclusive *ʔipah.* He is talking about general principles rather than the needs of the Batek. The cooling functions of the forest too are sacrificed.

Having returned the discussion to the outside world and what services we can derive from the forest, Tebu now circles back to his main theme:

We [inclusive] *can have a meeting. We meet, then we can discuss what to do. Discuss, decide, then we go ahead and do. Let's* [inclusive] *not give up the world. Let's not lose it. We* [inclusive] *should know how much to eat, how much to keep.*

His tone has become more urgent: *we can have a meeting . . . discuss what to do.* This is the key purpose of the message. Put aside the metaphors, the images, the cosmology: the political objective is to seek consultation. The Batek are willing to engage with the decision-makers; they want to contribute their ideas and communicate; they want to be heard. What he's offering is a scenario whereby Batek and the rest of the world sit down as equal partners in a meeting, face-to-face, discussing, deciding, acting upon mutually appropriate decisions. He reiterates why this is so important: from his point of view, we are ignorant. We need to be informed. We from outside the forest, the decision-makers and resource claimants of various sorts, are *giving up the world, losing it.* We do not know *how much to eat, how much to keep.* It is this ignorance, or this denial of proportions, that Tebu wishes to address.

He piles on the evidence:

Gob in thinking of roads will lay down oil palm. Consider that they kill the world. Where is everyone [inclusive] *going to live? So they kill dəɲaʔ heh* [our world]. *In the past, we lived healthy. Now, no longer can we* [inclusive] *want to be healthy. So everyone lives by common rules.*

Gob, the Malays, here standing for the powers-that-be, have a materialistic conception of the forest, for they are only thinking towards the roads and estates. When roads are built and forestlands converted, the forests cannot grow back. So the world dies: river flows are disrupted, floods will come, there's nothing to hold up the land. That Tebu uses at all the imagery of death is startling. It is probably not a traditional view on the world. This rhetorical development is consistent with a general trend towards integrating concepts of commodification and change with the environmental ideology (a point that we'll examine again in chapter 3). In the context of present-day Malaysia, Tebu is challenging commonsense notions of development. To many, having no roads is a symbol of being "left behind"—isolation, backwardness, deprivation. Hence, a common "gift" from politicians to communities at election time is the building or improvement of roads. As for plantation estates, Malaysia's success in cornering the oil palm market is often trumpeted as a triumph (p. 7). But what others mount as symbols of economic success Tebu characterizes as symbols of death.

But this is too easy. Tebu does not say that these things, roads, estates, township expansion, are intrinsically bad. Nor does he say "don't cut down any more trees." He does not advocate a retreat from affluence. What's bad is the overriding importance given to forest clearance, the sense that there is no attempt to consider other ways of "looking for food." As he said earlier, *let this be the limit.* When we only think towards material wealth or the programs that promote this, we "kill the world." The third-person pronoun throughout this passage is *heh*: you and us. The implication is global. For Tebu, it seems self-evident: it was healthy before, it no longer is. This restates his earlier commentary: we lived in peace before, we don't now. To him, what's happening to us today shows how we are all constrained by, subjected to—and ideally enjoined to observe—the same ecological principles (*hukum*: Malay for "order, command, judicial sentence, legal rules") as they, the Batek.

The skeptical reader may beg to differ. There is the overwhelming evidence (see pp. 6–7) of a dramatic rise in living standards in Malaysia. For many, the benefits of development are self-evident. When the Batek say the world is no longer healthy, are they perhaps making an unwarranted leap from their particular experience to the world at large? Even if we grant that environmental change has had deleterious effects on health conditions generally,[19] we can point to history. The millennia-old history of human exploitation of forests,[20] whether for timber or the so-called secondary products, was often accompanied by internecine conflicts and warfare at one level or another. We can argue that there has never been a "golden age" when anyone "lived healthy" or "in peace." For Malay peasants in Pahang certainly, there was a time in the past when "there was nothing to eat, there being a dearth of buffaloes, the planting of padi was difficult, and no one could be certain that he would not have to fly on the morrow."[21] From this perspective, the present is infinitely preferable. These are valid objections.

I would say, however, that such arguments miss the point. Neither Tebu nor any other Batek individual I know is innocent of history. His nephew and protégée Kayə?, for example, thinks back to the long history of slave-raiding in the Peninsula, when forest peoples were regarded as commodity (see pp. 102–8 for more discussion), and prefers the world of today. I think the fundamental problem that Tebu raises has to do with how we perceive and deal with environmental change. To accept degradation and its effects as inevitable by-products of development: or not to accept—to push for a questioning of development itself. Arguably, the scale of the impact means that choices are more constrained; as he stressed, we can no longer "*want* to be healthy" (emphasis mine). From the point of view of the forest, we may have reached what some writers call "criticality,"[22] the point of danger beyond which environmental changes are irreversible. Tebu wants us to accept that it is not only the forest or the forest people that are suffering.

Whether or not we accept the premise and take appropriate action, the Batek do:

We [exclusive] *miss the times of peace. We remember, we miss. We show how.*

That's their role, to "show how." His term of choice for this is *pə-hinĕk*, which means to do something so as to provide a "model" for a student to imitate or copy (*hinĕk*). Following on the theme of wanting to be consulted by the decision-makers, he means that he wants us to see how the Batek live, to understand their ideas and knowledge that we may become informed. In this particular passage, it is the "remembering" of the "times of peace" that we are asked to imitate. To *not* remember is to be blithely unaware of the effects of our actions, of how far on the road to degradation we have come.

Here is the irony of their position. They feel the worst effects of environmental degradation and therefore are best placed to remember what that environment is, at its best and its worst. The more one is entrenched in the built environment, shielded from the direct impacts of degradation, the easier it is to shrug off problems or, more abstractly, to put a fence between present and past. This luxury is denied to the Batek. But here Tebu suggests that this "deprivation" is a source of inspiration: their sentiments—always bittersweet in view of recent threats—are what enable them to know the forest and to use that knowledge to show others "how."

It seems evident that when Tebu says the Batek can "show how" he again has something more than the purely literal in mind. Because, in lieu of the face-to-face meeting he proposed earlier, he now switches topic and offers his insights, drawing from Batek ecological theory:

Already the earth is all cut up. The soul of the rivers is blocked. It's important to understand the danger. The rivers can no longer flow, they flood their banks.

Plate 3. Women and children at work and play on the Tahan River, Taman Negara, March 1996. As land and water resources degrade, so will the quality of such experiences.

The soil becomes lǝkɔc [soft], it collapses. They open up channels elsewhere, that's where the earth fissures.

Consider how systematically this concatenation of themes has unfolded. First we heard about the blasting of roadways (the auditory impact). That introduced us to the linkage with the other-world (Gubar, the thunder god) and the unpredictability of weather (rainfall patterns). Then we moved to something more substantial: we heard that the land is an island and how the trees stop the land from disintegrating. The topic of erosion, which we can define as the loss of matter from land, then immediately raises the question: what happens when that matter is moved out of its natural context? It creates, I would answer, pollution in another place. So we have been steered through two primary elements, air and land (including trees and forests), and we are now ready to move into the third, the waters.

I describe this unfolding of ideas in terms of a rhetorical strategy. But Tebu is also drawing from one of the most fundamental classifications in Batek thought, that between land and water. To maintain the order of the world involves maintaining the boundary between land and water: permitting rivers to flow their course, keeping the forestlands intact. Increasingly this boundary has been violated. I have already commented how hydroelectric dams symbolize the horror of inundation, the end of the world (see p. 22). In 1993, this was a common theme; the Batek often stressed to me then how they sensed more flooding events compared to the recent past. The fear of flooding is a common theme among Orang Asli, woven undoubtedly from the actual conditions of geography and meteorology. Historically, floods have been the major hazard in the Peninsula. Peninsula rivers exhibit a "dendritic" drainage pattern. Rivers branch off from each other and create a fine root-like effect on the topographical relief. Many tributaries will flow into a lowlying major river. As Ooi describes the geography: rivers tend to be "swift and narrow at their headwaters in the mountains and slow, sinuous and broad in the lower reaches. A sudden intensive fall of rain in the upper reaches might send a torrent downstream at a rate which the rivers at the lower courses could not adequately cope with."[23] That aside, the threat of unpredictable floods is exacerbated with increasing water diversion, such as those created by hydroelectric dam-building and forest clearance.

The word "soul" is a recurring theme, a leitmotif that strings together diverse passages and offers insight into the "hidden" meanings in Tebu's message. Earlier, he used the word to discuss the human impact of forest loss; now he resuscitates it to talk about the rivers. The soul of the rivers, he says, is blocked for it needs to flow naturally. "Block" is my translation from the Malay word that Tebu used, *sǝkat* (Malay *sekat*).[24] It captures the sense of flow being stopped, perhaps violently, by something physical. When the land is fragmented, the soil matter *ʔoʔlõʔ* (washes into the river) and the waterflow is blocked. With sedimentation, flowing rivers become ever shallower. The water has nowhere to go; it spills the banks, pushes through the earth, looking for alternate channels, making the ground ever softer, ever more fragmented, ever closer to final fissure.

How familiar, again, this all sounds. Take away the cultural specificities and the poetic imagery, and we are back in the mundane world of forest hydrology. From the point of view of ecology, none of these thoughts is at all extraordinary. But note that interjecting sentence: "it's important to understand the danger." The urgency has moved up a pitch, as it must when the prospect of inundation is ahead of us. Tebu continues at this level of intensity:

> *It's like, we* [inclusive] *look for food—from food we get rich, but the world is gone. We should know how to keep. It shouldn't be that we become rich and kill the world. Our lives become shortened, we don't live long, when we're too greedy. We* [exclusive] *know how to keep. When we [inclusive] make a living, we should value the soul of the world. But if they don't know to value the soul of the world, I don't know.*

This is a tray of oppositional choices that he presents to us. We can "make a living" and "become rich," or we can "keep" the world. To him, it seems that we cannot both be rich and not kill the world. This reiterates the earlier point that there must be a sense of proportion. Without proportion, we "kill the world." The "soul" of the world depends on recognizing this and being willing to assess which is more valuable: short-term wealth or long-term maintenance of the world, its rivers, its forests, its lands, its natural resources. If "they" (presumably the decision-makers) don't "know" what the values should be, there's nothing the Batek can do.

There is another meaning to this passage. From talking about the values that all of us (*heh*) should privilege, he inserts the observation that they (*ʔipah*) "know how to keep" the world. This joins up with the theme of knowledge and of teaching us what we need to be doing; *they* know the values and he wishes that *we* too know. But he realizes there are other forces at work: there are people who are blinkered by the search for wealth and they might be so ignorant as to devalue the soul of the world. Ideally they should be brought into this conceptual framework. See how all of this reverses the standard model of knowledge flow: it is not the Batek who are ignorant and need to be exposed to the right values. It is, rather, the "rich" whose consciousness needs to be raised, for upon *their* behavior turns the possibility of arresting environmental degradation. As many commentators have noted, it is often the "victims" of development who are made the scapegoats for environmental destruction.[25] Here the "victims" are challenging this, putting the morality of us, those in the metropolitan centers, under careful and critical scrutiny.

If Tebu cannot persuade us with his logic, perhaps he can—and this brings us to the final portion of this message—persuade us by example. He has given us a series of choices: how shall we behave, what should we do with the natural capital. In the following passages he makes clear what Batek have chosen and what that choice means for us. He concedes that this choice involves more than thinking about or using the environment in a particular way; it involves an entire way of life, a way of life based on mobility, whose logic may seem mystifying to outsiders, who often level the epithet "wild" at them:

> *We* [exclusive] *could be rich but we value the soul of the world. Now we've changed a little bit but we still remember the past. As long as we live in the forest, we'll give the instructions. We don't want to fight, to kill each other.*

> *There are Batek who want to be rich. It's not easy to live like this, we suffer. But this is preferable to killing the world, like life outside the forest. They bring about the end of our lives the way they live.*

> *We* [exclusive] *travel in one place then remember and go back there. People who live tame, they kill the world. JHEOA officers ridicule us for living wild. They don't know how to think. I want them to know how to reason.*

By way of wrapping up, Tebu said that other shamans had also wanted to talk to me. He was talking for them and the rest of the Batek too. They share his sense of worry. It is worth noting that the first-person pronoun *yεʔ* (I) appears just once, and that at the end: *I want them to know how to reason.* Until then, he was talking as a representative. Largely, that's how we must approach this message, as voices rather than *a* voice from the forest to the city.

At the start of this book, I wrote that its approach follows the direction laid down by the Batek. We can consider Tebu's intervention an orientation marker, pointing to the ideas that we need to examine, the themes that should branch off, the desired destinations. Like any good journey, there are also false trails that loop off into untenable positions.

I would say that Tebu's message crystalizes what many of the Batek feel, if not always so confidently or expressively. This is the reason for going step-by-step through the text of the message. The exercise has been revealing. It's allowed me to understand more intimately the verbal poetry in Tebu's persuasive techniques, his intelligence, his craftsmanship. It certainly brings to the fore aspects of his character and thinking that I'd not hitherto recognized.

He began with a straightforward declaration of the *problem* (what happens when the forest disappears; what the Batek want the world to know), followed by supporting *evidence* (auditory and meteorological disruptions; road-building; plantation advances; the decline of health).

Quickly, he moves on to what we should *do* (balance the search for wealth with knowledge of what to keep and save; consult with the Batek): the question of our *responsibility*.

Failure to meet this responsibility will have far-reaching *consequences* (threaten the lives of the forest people; threaten our own future as the land collapses around us).

He nudges us in the right direction by reminding us of a qualitatively better world than exists today: a call for the importance of *memory* or, if you prefer, a plea for historical consciousness. This point primes us to consider nonmaterial dimensions of environmental destruction; the assaults on the world's "life," "heart," "soul," "veins and tendons"; the creation of *loss*, a loss felt keenly by forest peoples who have not discarded their links with the past.

All of which was enclosed within a cosmological frame: the science of how the world originated and developed. Through some key concepts, we were introduced to the language that the Batek can use to talk about the ecology. Each of these concepts was a signpost along Tebu's route: noise, sky, trees, forest, land, rivers. "Soul" was a recurring motif: soul of the forest people's lives, soul of the rivers, soul of the world. The intangible, the essential: that which defines the nature of a thing or person, that keeps it alive, that is threatened by present levels of resource exploitation. Underlying the cultural idioms, we detected much affinity between Batek explanation of natural process and ecological science, a point that will appear again in this book. One aspect that's different is their attribution of willful agency or intentionality to other-worldly forces. There is no rigid separation between the world of "this" (ecology; materiality) and the world of "that" (spiritual); these are parallel worlds that mutually respond to one another's actions. As such, it may not be correct to make these distinctions in the first place. Tebu is commenting pragmatically on the causes and effects of environmental degradation as he observes and perceives them and it so happens that part of his explanation falls within a domain that *we* might call spiritual.

At every stage, as marked by different uses of the pronouns *heh* and *ʔipah*, we were reminded of *relational* unities and tensions. Sometimes this involves a tension of opposites: between "them" and "us," between their way of life and ours, between their remembering and our forgetting, between their wanting the world and our killing it. But Tebu's central objective is to show how these differences pale before the reality: without the forest, there will be no world. The effects of forest degradation are felt already. The forest is the skeleton that holds up the body of the land and without the forest the land collapses and the world will disappear. Where, in asking me to regard their stories as *surat* (letters), the Batek forced me to reckon with their need to communicate with the urban world and my part in that communication, Tebu in this communication forces us to recognize our relationship to the forest and its peoples, and the relationship between our behavior and the future of the environment.

Tebu does seem to hark back to a "golden age," before the advance of worry, insecurity, and threat of destruction. The past provides a point of comparison for the present and future. He's taking from and building upon a tradition of thought. By the standards of his community he is not a rebel. But tradition in the Batek sense is always problematic. If, as I stressed in chapter 1, they have a history of integrating external ideas and resources into their repertoire, so in some sense are they open to *change*. To express this differently, *the more they remain the same the more they are changing*. Thus we do not hear Tebu voice a resistance to development or a wish that outsiders would leave the forest alone: such is not the Batek way. What he critiques is a headlong tumble into environmental destruction via the clearing of forestlands. Nor should his nostalgia be dismissed as an Arcadian search for the impossible. The question of moderation and restraining the search for wealth is what he wants us to consider as alternative modes of being. Doing that means changing environmental behavior: using up as much as we can while we can or keeping in reserve for the long-term maintenance of the world.

All this is honed within a political position. As the message implies, they do not have the authority to prevent degradation but they have the knowledge of its effects, ramifications, and implications. It is their very lack of political power that puts them in that position and therefore, whether they have chosen it or not, makes of them authorities on degradation. He outlines three ways that the Batek can "show how." The most indirect is behavioral: so long as they continue to live in the forest, guided by its moral constraints, they will provide a model for our edification. More direct is oral exposition: in this message, he outlines what he wants us to know about the importance of the forest to the future. Even more direct is face-to-face talking among Batek and decision-makers, a scenario that at present remains abstract and quite distant.

Here is a challenge to universalizing divisions between modernity and tradition, urban and rural, progressive and primitive. In Tebu's message it is not the primitives ("the wild people") who lack knowledge and awareness. Rather, the progressive sector ("those who live tame") is "killing the world" and displaying dangerous ignorance. The latter have the authority to halt degradation but they're not attending to the Batek. Hence the urgency of communicating their findings. In so doing, they flip over the classic assumption that it is the "primitives" who need to be "invited"[26] to join the state, whose interests need to be "represented"[27] by others. Here it is the "primitives" who are proactively inviting the state to join them and take "instructions" from them. Before that invitation can be taken up and appreciated on its own terms, we will need to change how we perceive the forest and the forest people.

The Approach

The perspective of this book is clearly biased in favor of the Batek. As instigated in 1993, my job is to deliver the "letter" entrusted to me. In so doing, I hope that this study will add a relatively unknown perspective to current environmental debates. I intend to honor Tebu's—and by extension the Batek's—wishes to be heard. Their dwelling place is run by government agencies; even their very identity is constructed as primitive and insignificant to the broader Malaysian society. This is like the creation of nonexistence. They challenge that. They feel that they have something to offer to us. That is a rare invitation: to see the world through the eyes of a people who mainly appear in public discourse as tourist objects. As such, though this book is principally concerned with the Batek, I hope that it provides a mirror effect, provoking readers to question self through knowledge of the other.

At heart, it is in the classic mold of descriptive ethnography. While I could have chosen to follow a well-trodden anthropological path of focusing on "discourse"—the production of speech and text—I believe there is much greater value in taking readers *into* the forest: to know the people there, what they do and say, how they relate to the world, what are the constituents of their world, how environmental degradation affects them, what meanings they put on degradation. There is no end of studies on issues like extinction, land degradation, and forest cover change. Not many will come to know the Batek. And thus, my objectives are: further along our understanding of Tebu's ideas and the political position upon which they rest and bring to the fore the values that the Batek attribute to the forest.

In looking for an "angle" from which to tell this story, I focus on two interwoven themes that crosscut Tebu's major concerns: *place* and *loss*. The main gap to be addressed is intellectual or cognitive: the widespread tendency to marginalize and simplify the forest so that it is felt to be (variously) a source of wealth, an obstacle to development, a wilderness inhabited by savages. Hence the bias in favor of the Batek. I do not necessarily value preserving forests as *relics*: intact reminders of the past, with biodiversity that potentially bestows undreamt-of wealth to those who know how to prospect for its genetic resources. That is not the forest that will come through in this book.

I begin in chapter 3 with categories of place, those involving: world, land, forest, land types, vegetation, rivers, animals, and plants. This is a general account of the geographical setting. What enables someone to become intimately familiar with this place and all that it contains is the moving body: as people journey along the pathways, from camp to forest, from camp to camp. With degradation, as the landscape both within and outside the forest changes, so too perceptions. Concepts that were hitherto implicit become ascendant (and vice versa) and thus lies the practical basis for an emerging environmentalism. While individual perceptions may be variable, everyone draws from a common stock of principles and norms, which may be elicited from the stories that people tell each other. And so chapter 4 goes back in time to the moments of creation, when the landscape assumed its primordial shapes and forms. This examination of the origin myths—a form of oral history—flips over the question of loss by asking: how did this abundant world come to be and what was it meant to be? The Batek's rich narrative tradition shows no sign of dying out and in telling—and therefore remembering and reproducing—the stories, so do they hold up an ideal from which to compare the present. I suggest that ideals of this type in some sense enable people to construct a morality of change: i.e., a body of principles that says what kinds of change are and are not acceptable.

Other kinds of stories abound and chapter 5 turns to narratives of place: not of the ideal place as represented in the origin myths but a place of degradation. Places have biographical, temporal, and sentimental associations. Memories are inscribed in a place or, to put this differently, place is evocative of history. Most important for the preservation of environmental memories is having the place to return to. As landscape deteriorates and familiar places are transformed or appropriated by outsiders, it may be harder to return and recapture the meaning behind the old stories. Landscape knowledge is reduced to cognitive activity, no longer enriched by experience and ongoing discoveries. The second part of this chapter then examines—in somewhat broad strokes: what does it mean to belong to this place? Rather than focus on the positive ideals, a story that probably only the Batek could tell fully, I highlight their sense of vulnerability for this, too, is integral to the sense of place. For that purpose, we begin historically with a synoptic view of external exploitation of the forests, its products and peoples. This history has cultivated a strong sense of being at the butt-end of outside schemes and projects. Ideas of predation are condensed in two powerful symbols: Malays and tigers. Both are on the edges of the in-group: belonging and yet not-belonging. From this examination of internal/external and human/animal relations, a larger view is obtained of the society of the forest: those members that belong and don't belong and what are the principles of relatedness.

Chapter 6 then moves away from stories, towards the biological richness of this place and, more abstractly, the use values of the forest. It examines how the Batek know and exploit two key vegetal resources: wild yams and seasonal fruits. In relation to the broader themes of this book, this chapter reminds us that not all is lost: much persists and the forest continues to be an important source of sustenance. The question is what will happen if access to these resources were lost. Further, I hope to show how much this "light-impact" foraging strategy—opportunistic, variable, and spatially extensive—alters and in turn is altered by the forest ecosystem. External perceptions of the forest as a wilderness inhabited by wild peoples eating wild foods come to seem most erroneous.

The collateral theme of knowledge appears in various guises in the early chapters. In chapter 6, resource knowledge comes to the fore. Accordingly, chapter 7 then rounds up the picture with a tentative examination of perceptual knowledge, as signaled by the images that are seen and the noises that are heard in this place. As Casey remarks, "Perception at the primary level is synesthetic—an affair of the whole body sensing and moving."[28] Thus ends this final ethnographic chapter with a return to the pathways and camps and an argument for why movement—so often denounced as "nomadism"—is integral to the knowing of place. The conceptual objective is to draw attention to different ways of knowing the forest and, reciprocally, the forest's role in reproducing history, knowledge, and knowledge of self and community.

The conclusion reflects on the title of this book: changing pathways. It is deliberately ambiguous. Who should change? What are the pathways? What are the mechanisms preventing change? What are the prospects of improving communication between the Batek and the state?

There are eight maps and three appendices in this book as well as a glossary listing the key Batek terms that appear in the text. As promised, Tebu's message is presented in appendix A, which reproduces my original transcript, followed by a word-for-word translation and the "free translation" that was examined in this chapter. Appendix B is a transcript with translation of a route description that shows how landscape categories are applied in giving directions. Appendix C addresses chapter 6's data on wild yams; it presents Batek notes on each of the recognized wild yam species. The major place-names (toponyms and hydronyms) mentioned in the text are defined or spatially located in the glossary. Batek individuals whose names appear in the text are listed in the index.

A subsidiary theme in this book is to challenge external constructions of the forest and the forest people, the effects of those constructions, and, more broadly, ideas of "nature." In Tebu's closing passage, we heard his critique of "tame" people: as chapter 5 will show, the "wild" and "tame" distinction is inherited from some well-sedimented Malay folk ideas about Orang Asli. At some level the Malay ideas parallel the Hobbesian view of nature and "people of nature." The Batek challenge is consistent with that found among many indigenous peoples around the world: that the distancing of nature is somewhat suspect, that the distinction between culture and nature is so fuzzy as to lack any meaning, and therefore there is no "nature" set apart from society. However, "nature" has its ambiguous qualities and evokes ambivalent responses among the Batek. It is hard to think of them as "living in harmony with nature." Accordingly, this book contributes to a broader questioning of such images and advances a more nuanced interpretation of hunter-gatherer conceptions of the environment.

Tebu's willingness to speak on behalf of others shows that he is trusted to deliver. I sense a risk here in focusing too much on him. There are as many perspectives on the forest as there are Batek. Different social interests abound. But: three things suggest that Tebu accurately expresses general concerns.

One, as I stated in concluding my exegesis of his message (pp. 39–40), what he articulates is really a consensus statement crafted through many conversations with other Batek that I was not privy to. Again I stress: the interview context was wholly spontaneous as far as I was concerned. Were it up to me, I would not have chosen that night, November 4, 1996, to have this discussion: two nights before leaving the field, my mind already halfway out of the forest, distracted by the emotions of saying farewell to people who had been my closest companions for over a year. I have not seen Tebu since 1996; he was not around on my flying visits in 1998, 1999, 2001, and 2003. I kick myself violently for not pursuing issues more deeply in those final two days but I was not in the right emotional state for it. The second mitigating factor, then, is that I did not set out to find a message like Tebu's. I did not impose my ideas on him. That was far from my mind.

The third factor is that Batek social conventions—which have their core in the egalitarian ethos—inhibit people from freely representing others. Anyone who tries to stand out from the crowd, to advance an agenda that is not compatible with general views, would be ridiculed, perhaps ostracized. Tebu took a political risk in addressing me; but it was a calculated risk. For one thing, he had known me since my initial appearance in 1993. I was a "safe bet." For another, he would have known that he was not speaking out of turn, that many other individuals would agree with his sentiments. Although few would be able or willing to express themselves in his style, I think he was right.

Nevertheless, in order to get as full a picture as possible, he is not the main source of information for this book. I draw on a varied body of data: observations of the landscape, documentation and records of behavior, materials from language, myth, folklore, ritual, and belief, and what different individuals have said both in and out of the awkward setting of interviews. The discussion *is* structured as a conversation with Tebu, an extended engagement with the implications of his ideas and an elaboration of themes introduced by him in this chapter. I cross-reference to him frequently in introducing each of the major themes. The approach is therefore inductive.

My preference is to emphasize what people say. But I also insert much interpretation of my own. I believe that it's possible and even important to offer judicious cultural interpretations—especially of aspects of symbolic meaning and spiritual knowledge—and wherever necessary I do so. However, in view of the romantic appeal of indigenous peoples today (as exemplified in the common tendency among urban-industrial peoples to turn to indigenous cosmologies as a foil to market-driven paradigms), overinterpretation carries the risk of inviting caricature. My fallback is the empirical approach: evidence, the kind of objective, observational evidence that Tebu offered in his message. I tell stories. What I've seen, heard, recorded, read, thought.

To make it clear when the Batek are speaking and when I'm interpreting on their behalf, I stress who said what, where, and when; attributory information like personal names, interview dates, and place-names abound. Where I'm confident that an idea or concept is oft-reiterated or socially shared, the generalized "the Batek" will do. Detailing interview contexts has the advantage of "personalizing" the book, highlighting individual takes on ideas and issues. I hope that Tebu's voice is not the only one rising above the background chorus. This does privilege the perspectives of *some* individuals (those I was closest to) but it is not culturally inappropriate. The Batek themselves often talk in that way: so-and-so said that, that's so-and-so's song, etc.

Admittedly, this book is a synthetic presentation that papers over internal dissensions. I am aware of and agree with the argument that studies of local knowledge privilege some voices over others. It can be shown that underneath the apparent conformity there is tension and disagreement, and one can expect that the promise of a comfortable life outside the forest, a life based on a more intensive use of natural resources, would have some attraction even for the most culturally conservative of the Batek. As Tebu said, *there are Batek who want to be rich.* As much as the Batek are looking out of the forest, they're also looking around at each other, observing and judging behavior, advising against going in one direction rather than another. A more sociologically nuanced study of this process (the micropolitics of community life) has its value—*elsewhere*. Among the Batek individual silence is usually a strategic choice. It is the larger silencing of the *Batek voice* that should most concern us here.

Notes

1. Nicholas 2000, 192.
2. Ellen 1999, 152.
3. Ellen 1999, 148; italics in original.
4. Brosius 1997.
5. Brosius 1997, 476.
6. Brosius 1997, 474–75.
7. Conklin and Graham 1995.
8. Hood 1990, 148.
9. Jomo 1992, v.
10. See Vogt's (1995) attempt to set the record straight. For external reviews of the events, see Dentan et al. 1997, 77–78; Ismail 1995.
11. Kirk Endicott 1979a, 131.
12. Lye 1997, 2002, 2004.
13. I thank Mano Maniam for this insight.
14. Lakoff and Johnson 1980, 5.
15. Ingold 1996, 133, citing Michael Jackson.
16. Pillay 1996, 432.
17. For a more general discussion of the superhuman beings, see Kirk Endicott 1979a, especially pp. 124–27.
18. Kirk Endicott 1979a, 124.
19. See, for example, Pillay 1996 for a more general overview of Malaysian environmental health.
20. Dunn 1975.
21. Gullick 1988, 29, citing words spoken in 1892.
22. Brookfield et al. 1997.
23. Ooi 1963, 252–53.
24. Though there are many Batek words to describe the conditions of river, rainfall, and water, or to describe how different objects touch water (e.g., whether they fall in, are

dropped in, or slip in), to my knowledge there is none for the physical blocking of water flow, such as by soil matter, as here, or by built structures like dams and walls.

25. See, for example, Dove 1983.
26. Tsing 1993.
27. Nicholas 2000.
28. Casey 1996, 18.

Boys posing for one of their friends. Anonymous Batek youth, 1996.

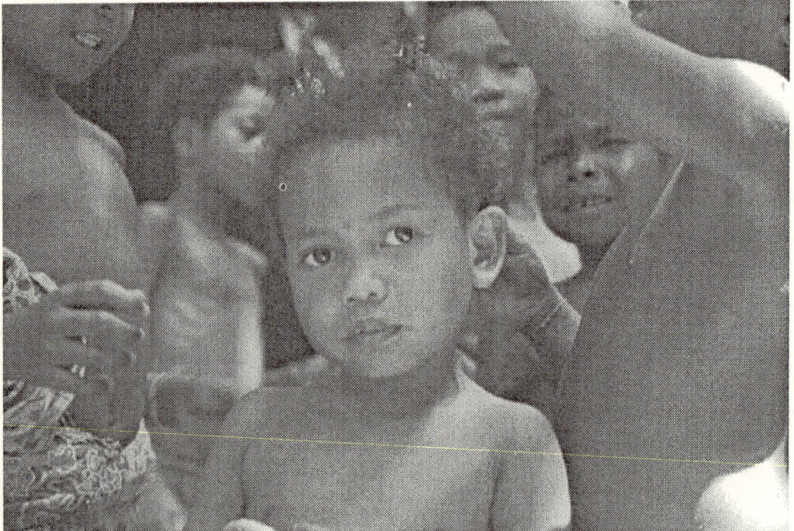

Little Diboh (front) is festooned with hair clips given by tourists, 1999.

Plate 4. Portraits of Batek children

3

The World of the Forest

The Batek probably first sighted low-flying airplanes around the early 1950s, at the height of the Emergency. Until then, the *həp* or forest must have seemed quite secure. When there was danger in one place, the Batek could escape to another. The forest was an effective refuge. Now, there was aerial bombardment. The refuge had become vulnerable. Others were fighting a war. The Batek couldn't have known when to expect the next fleet of planes in the sky. And then, the bombardments stopped. In 1989, the Communist Party of Malaya officially surrendered and disbanded; when I met the Batek in 1993, they knew the "*komini*" days were over. They sometimes recall the stories. There was one about Semai women who had been kidnapped by *komini* and asked Batek to help them. Or about being approached by *komini* and running away to avoid getting involved. Most gravely, they recall the bombardments. They could shelter in caves—if they could reach them in time. Bodies were injured; children were lost; the feeble and infirm were left behind. Some Batek youths today hoot with laughter when they hear of their ancestors' terror to see those early planes. They cannot imagine what it felt like.

Batek experiences, and their ideas of their place in the forest, are not the stuff of newspaper headlines. During the Emergency, for example, they were not in the frontlines of battle. They were not among those Orang Asli resettled for security purposes or identified as giving food and aid to the communist insurgents. They have escaped this dubious fame. But their ideas of the forest have had a quiet history of their own: informed yet not determined by external events. Events like aerial bombardment, and other kinds of invasions by alien forces, cannot have left the Batek unmoved. The English word "bomb" has entered Batek language (as *bom*) and, as we've seen (p. 28), now describes any use of explosives. Further, as introduced at the start (pp. 16–17), they have a habit of taking ideas from here and there, inside and outside the forest, putting them together, linking up causes and effects, and coming up with "something" that is recognizably their own take on things.

This chapter describes their overall representations of the forest. It is a general account of the geographical setting as perceived by them. My approach is to run through some socially shared sets of terms and concepts, each of which offers a different perspective on the important components of the landscape. More broadly, this chapter introduces ideas about "nature" and in the latter half suggests how those concepts might change in response to degradation and threat.

The forest is integral to the Batek's sense of self; after all, they call themselves *batɛk hǝp* (people of the forest) and they consider the forest *tǝmpɛt ʔipah gǝs* (place where we dwell). Yet the forest can also be distanced conceptually: as people of the forest, Batek say, they *jagaʔ hǝp* (guard the forest). They perform this function, not anyone whose home is outside the forest. And if there were no people in the forest, the world would be destroyed. Not only is the world's continued existence dependent on the forest, that forest has an ethnic identity: defined by Batek norms, populated by Batek.

Spatially, the forest is the major frame of reference. It can be all around one while walking therein; in the foreground, right in front of the eyes as in a small camp; or in the background, as from the perspective of villages and towns. On an everyday level, it provides practical materials, nourishment, and protection, a sense of community and history, and is central to the Batek's construction of their identity and ethnicity. It is where one's friends and relatives live; children's playground and schoolroom; a place to walk in; to go visiting in; stocked with an abundance of useful resources. It is, among its multiple uses, a source of intellectual sustenance. The pursuit and gathering of knowledge is a deeply satisfying source of meaning. As Kirk Endicott aptly expresses it, the Batek "identify closely" with the forest, regard it "as their true home," and "consider their living in the forest to be part of the natural order of things as established by the superhuman beings."[1]

Though this image is unremittingly rosy (a point I take up more fully in the next chapter), the forest also harbors a number of entities, human and nonhuman, that have the power to unleash danger, malevolence, and take human life. The most dramatic of these is the thunder-god, Gubar, who featured early in Tebu's message (see pp. 27–28). The Gubar phenomena is part of the famed "thunder complex" in the region.[2] Batek often interpret his wrath (thunderstorms) as a comment on moral conditions in camp and a response to certain misbehaviors. Their philosophy strongly emphasizes the maintenance of proper relations in the social world (the social here also incorporating the nonhuman persons) so as to avert the destruction wrought by malevolent forces like him.

Still, though the forest has this ambiguous quality, it remains the safest refuge to hide (*ʔɔt*) in. It enables the Batek to escape external control. Withdrawing farther into the forest, drawing on their unparalleled knowledge of its topography and walking routes, remains their major mode of resolving disagreements with others (for example, irate employers, forest-edge villagers, or government administrators). Withdrawal is a survival strategy that evades control by outsiders, preserves political autonomy, and protects physical freedom. However, it feeds into the prejudice that the Batek are free-roaming nomads without a developed sense of civilization or place (see pp. 96–97).[3] In the critics' admission of themselves as people who could never follow, let alone track down the Batek (a sentiment often expressed to me or in my presence), this prejudice also reveals much about outsiders' failure, conceptually or practically, to "penetrate" the forest.

For clarification purposes, all writings in this vein must synthesize and simplify and will not fully capture all the subtleties in Batek knowledge and perception of the forest. First there is the issue of perspective. People can take different perspectives on the forest depending on what they're doing, where they are, where they're going, and who lives there. A bigger obstacle to an outsider is the nature of knowledge itself: there is no overarching definition of the forest, set down in explicit language, that every child grows up learning to enunciate. Accordingly, the Batek do not put in so many words what *həp* is. Perhaps that is not necessary to them. While the word does have the straightforward meaning of "forest," theories of the forest are not systematically represented. However, it would be wrong to say that there is no conceptualization of *həp*. After all, the thrust of Tebu's perception of degradation is that something is being lost and replaced with something else. Underlying that anxiety is a concept of what those "somethings" are and how they compare with each other. How to get at them? At the concrete level of environmental images. Examining some basic idioms, ideals, and characteristics will give us fragments of definition here and there. Which correspond, perhaps, to the way the forest is perceived: situationally and incrementally.

World, Land, and Forest

We encountered a few central concepts in the previous chapter. Let's start with the major one: *dəɲaʔ* (world). Its meaning is the most inclusive of all. A Malay loan (Malay *dunia*), this word encompasses *everything* in the world. The *dəɲaʔ* contains within it all the integral landscape (phenomenal) features: those that stand above the ground, the ground, the rivers, and the watery world below. A close reading of its usage by Tebu may be useful here. Statistically, it appears thirteen times in his message (see appendix A); it is the most iterated landscape term there. In the context of the message, it seems clear that it encompasses the world beyond the forest. Overwhelmingly, the *dəɲaʔ* is generalized. In every other usage Tebu signifies possession with the inclusive pronoun *heh*; i.e., *dəɲaʔ heh* (your world and our world). It is the world that everyone lives in.

What are its characteristics? Tebu's language provides some clues and it is perhaps telling that the words he uses are all Malay loans *that have Batek equivalents*: *habis* (end; Batek *jaʔ* or *nen dah*), *bunoh* (to kill; Batek *sakɛl*), *ɲawaʔ* (soul; Batek *rway*). Three of *dəɲaʔ*'s appearances in the message are together with the word *ɲawaʔ* (see appendix A: paragraphs 18 and 19) while *bunoh* appears four times (paragraphs 15, 18, 19, 21). The life of the world depends on maintaining the underlying relationships among all that it contains, particularly forests, lands, water bodies, vegetation, fauna, and geomorphological forms. The direct causes of its ending are: forest clearance, road-building, plantation establishment, and overexploitation of resources. Thus the world has both noumenal *and* phenomenal components, which are integrally related. What links these is what people do to the world; the world can be "killed" through anthropogenic activities, and the decline of health and peace are attributed to the assaults upon its soul.

We do not know when the Malay words entered the Batek lexicon, under what circumstances, and for what purpose. The use of *ɲawaʔ* and *dəɲaʔ* are certainly long-established. Tebu's usage of these words tells us two things: verbal adaptability (using a parallel language to make ideas more accessible to the intended audience, since Malay is the lingua franca in Malaysia) and the emergence of a newer type of environmental discourse. I've commented on the first already (pp. 29, 31). On the second, Tebu's word choices are absolutely consistent: put all the words together and we get a picture of the *dəɲaʔ* as a living thing, endowed with soul-stuff and subject to all the processes that might snuff out a life: exploitation and murder.

The imagery of death is startling; it is not what we would expect to find among a people romanticized as primitive nomads left behind in time. Which marks a shift in environmental perception. In the past, before the onset of degradation, possibly the Batek did not have an acute sense of the finitude of the world. It would not have been important to know whether the world lives or not. It just *was*. That is the image conveyed by Kirk Endicott's study of the religion, based on early 1970s fieldwork in Kelantan. Now, however, the future life and vitality of the world has come up for question. A language must be developed to talk about these newer conditions. Somewhat ironically the only available language is that of the people most identified with acts of degradation, the Malays, the ethnic and economic majority in Malaysia.

Take *bunoh*, for example. The Batek equivalent *sakɛl* has the highly specific meaning of murdering a *person*. Until recent times, there would have been no context for imagining how the *world* could be killed. Now there is: "*ʔayaŋ heh kayaʔ heh bunoh dəɲaʔ* [It shouldn't be that we become rich and kill the world]." There is now indisputable evidence of socioeconomic disparities in Malaysia (pp. 6–7), of the costs incurred by the behavior of some sectors of society, and the Batek must think through the implications for the forest and for them.

Metaphysics aside, threats to the world can be viewed in commercial terms. The world is not without tension. It can be perceived as an item of exchange, something that can be held back, given away, subject to struggles over control; it can be "*rugiʔ*" and "*gəh*" (appendix A: paragraph 14). These words appear close together in a passage where Tebu is urging us, his audience, to collaborate with them: "*Heh gəh dəɲaʔ. Jaɲan heh rugiʔ dəɲaʔ*" The most familiar connotations of both words relate to the circulation of material objects. *Rugiʔ*, of Sanskrit origin, is most likely borrowed via Malay (*rugi*); it is a common Malaysian epithet to describe "loss" or non-physical "injury." When someone *rugi*, that means he's lost out: either been at the weaker end of a deal or lost the opportunity to profit.

Gəh, a Batek word, is used in material transactions (the giving and taking of things): its straightforward meaning is "don't want to give." Though the original meaning of *dəɲaʔ* is abstract—the world in which we live—it is perceived as having value that is independent of it, that can be conferred or withdrawn.

Which suggests that there are at least two sides to this transaction. One side should stop the world from ending, should refuse to surrender it, and another side (those who "kill the world") would try to take it and therefore make everyone else lose out. Tebu's use of these words is strategic; he puts us on their side, who would surely not want to lose out and *will*, if the other party assumes full control. In the context of the environmental paradigm, these words acknowledge that a "price" has been put on the world. That there is a contesting ideology out there, powerful, threatening, morally questionable.

All this underscores the world's vulnerability. Other than the forest, another vulnerable part of the *dəɲaʔ* is the *teʔ*, which is likely to be the main geographical category. *Teʔ* does not appear in Tebu's message as often as *dəɲaʔ* (five instances total); its meaning is more localized and less inclusive. Its characteristics, as outlined by Tebu, are that it needs the trees to *tahan* (hold together or hold up; appendix A: paragraph 6); it has a heart (paragraph 8); it can be softened, broken up, fragmented, and it can collapse (paragraph 17). The relevant expressions in this respect are: *jagaʔ teʔ* (to guard the land) and *teʔ neŋ uhm tahan* (the land is unable to hold up). As with the world, then, land can be protected or destroyed through human activities.

Teʔ's best definition is "land" or "all the land" (that is, "the earth") but it can also have the concrete meanings of "soil" (the physical stuff) and "ground." As we'll see below, it is also a landscape classifier (i.e., an ecological category) and appears in a few topographical categories. The idiomatic expression for "widely traveled" is *teʔ tɔm hal* (lit. land river tracks); i.e., to be widely traveled is to make tracks all over the lands and rivers. Thus it is the surface on which we leave the traces of our being and it contrasts with the water bodies. *Teʔ* is also an orientational term, in which respect its connotations are always ground-centered (terrestrial). Here it contrasts with the sky and the celestial world and any upraised surface: when the ritually important annual fruits appear (see chapter 6), they are said to *cital bah-teʔ* ("to jump earthwards"); to sit on the ground (rather than in a house or lean-to) is to *ŋɔk hat-teʔ*.

Like the world, land *seems* to go on forever (but see below). It is as old as the world and perhaps even older than the forest. Land does not have a socially organized boundary. This is probably because the Batek are not systematic land managers, unlike (say) shifting cultivators, and do not reserve different productive activities for different types of land. Politically, they do not recognize individual ownership of land. Thus, they never identify any parcel of land as *teʔ Batek* (Batek land) or *həp Batek* (Batek forest). However, it seems that they do recognize that such a designation might be possible. They will point out if a parcel of land is *teʔ gob* (Malay land), thus suggesting that, apart from its inherent qualities, *teʔ* may also be identified with the way of life characteristic to a place.

There is no confusion between *teʔ* and *həp*. Where *teʔ* refers to the terrestrial realm—the ground beneath the feet, as it were—*həp* refers to the forest vegetation that grows upon it. *Teʔ* therefore has a larger surface area, and encompasses more topographical features, than *həp*. Recall Tebu's warning, however, that ultimately the land is an island and therefore finite: the forest rests upon it as it in turn rests upon the watery underground. Both people and the forest could not exist without the land, as it is the surface that keeps the underground waters from welling up. In relation to land and world, the forest is smaller. What make it important to the soul of the world are the ecological relationships: the forest firms up and maintains the stability of the land. To *jagaʔ teʔ* is to watch over these relationships. Together, the *teʔ* and the *həp* are the central features of the *dəɲaʔ*.

Landscape Concepts

Contained within the *həp* are its landscape features. One way to get a rapid view is to identify what the Batek consider salient there.

Let's start with those objective features that are also recognized as identity markers and cultural symbols, which therefore can be used also to mark territory. The *hayãʔ* (lean-tos; camps) and *tɔm* (rivers), together with the network of *halbəw* (pathways), must rank as the most important cultural symbols. They are critical to the Batek's understanding of who and what they are. We could say that these pathways, rivers, and campsites mark out the basic geography of the landscape and form the spatial framework around which the Batek organize their environmental knowledge.

Salience also shows how the perceiver recognizes natural discontinuities in the land. Topographic categories are certainly part of the image of the forest. The terms mentioned below are often referred to the problems of movement (i.e., of *jok* [to travel from one place to another]). As an example, I present in appendix B a route description from Lus, which shows how landscape features are used in giving directions.

The Batek recognize different categories of forest formations (I follow Kirk Endicott's definitions here as his explanations are clearer than those I got from the Pahang Batek): *həp ləy* (standard lowland forest); *teʔ təraɲ* (hill forest); *teʔ barɔs* (limestone soil forest); *bəlukɛr* (secondary forest); *teʔ laɲeh* (mountain or high ridge forest). The labels indicate that distinctions are based on topography and forest cycles. While *həp ləy* and *bəlukɛr* are types of forest growth, *teʔ təraɲ*, *teʔ barɔs*, and *teʔ laɲeh* are types of land. The meanings of the secondary lexemes in the labels (*ləy, təraɲ, barɔs*, and *laɲeh*) remain mysterious.

According to Endicott, *həp ləy* is "the general, unmarked category with which the others are compared."[4] It includes both disturbed and undisturbed forest. It is recognized (by ʔeyPaliy, for example) as the easiest kind of forest to travel in, simply because elephants also use *həp ləy* and they open up passageways that people can then use. Less desirable to the Batek is *teʔ təraɲ*: "Awful. I never want to *jok* there. It's full of thorns and hills" (ʔeyKapey). Apparently the trails in this type of land tend to be overgrown, so it is easiest to lose one's way there.

In marked opposition to *həp ləy* is *teʔ laŋeh*, which is like the outer frontiers of settlement for the Batek, who do not live in highland forests. It has often been remarked that Semang historically "are not mountaineers, or even hillmen. They like river valleys; at the most foot-hills; formerly even coast lands."[5] In addition to montane forestlands, *teʔ laŋeh* also represents any belt of land rising from the upper slopes of the foothills.

If *teʔ təraɲ* is associated with getting lost, *teʔ laŋeh* is associated with the sheer problem of finding a throughroute. ʔeyPaliy again: "it's hardest to travel in *teʔ laŋeh* because when you come to the end of a river, you'd have to look for the next river. That river would be on a different ridge altogether." They are useful as shortcuts, however; one can follow rivers to the source and then crest over to the other side (*linaŋ* or *rilɛk*) and enter the watershed of another river, thus avoiding the more circuitous routes in *həp ləy*.

Teʔ laŋeh is often avoided for another reason: the terrain is more broken, there are more high ridges, and it is therefore more strenuous to walk on them. At the transitional point, neither lowland nor true montane forest, food sources may still be found. However, most of the wild yam species mentioned as occurring in this habitat (*kəbaʔ, takop, kənsey, payol, tãw*)[6] are common in *həp ləy*, so there is no real need to go into *teʔ laŋeh* to find them. As Endicott continues, "they tend not to camp on hilltops and ridges because of the scarcity of food and water, but they sometimes enter *te' langeh* when traveling or hunting (especially for gibbons and siamang)."[7]

More often mentioned in conversation are those parts of the landscape that "stand out" and serve as landmarks. These are regular, meaning not unusual, features of the landscape but they are relatively distinctive and easily remembered. They can be used to orientate movement and give directions. Examples are: *ray* (lightning scar in a forest patch), *batuʔ cənɛl* (lit. mythicized rocks: i.e., rock formations whose origins are explained in etiological myths), and *padaŋ* (single species grove).

Among these, perhaps the most significant as landmarks are the rock formations: rockfaces, outcrops, cliff overhangs, and stone pillars. These are uninhabited but distinct components of the landscape, to which names may be given. People feel that these landforms belong in and to their landscape. These formations are not widespread. Topographical maps of the Lipis district, for example, show an even spread of lowland dipterocarp forest, dotted here and there by isolated outcrops. Up in the Kenyam valley (Taman Negara), there are some ancient stone pillars that the Batek consider of great ritual and cultural significance; these, too, are integral parts of the *dəɲaʔ*.

Batuʔ cənɛl is a special category of rock formation; they have supernatural origins that are explained in myths or folklore. Telling stories about (or to explain the origins of) land formations is of course a common thing to do, with long-term residents generally possessing the most detailed of the lore. In Southeast Asia, as with the Batek, many rock formations are said to be the petrified remains of people or communities that did not observe proper rules of conduct: notably, those beliefs and practices associated with the thunder complex, introduced earlier (p. 50). The existence of the rocks is therefore considered to verify the stories and validate the beliefs. So other than their landmark value, these rock formations also serve a moral purpose, reminding people of the consequences of misbehavior. They may serve as territorial markers but they also communicate a sense of place—the psychological certainty that one belongs to this landscape because one knows the stories of its origins.

But there are more ephemeral qualities of the forest, which also have their saliency. Treefalls are one source of ephemera. It is not uncommon to walk through the forest after a thunderstorm and find that a tree has keeled over. When the fallen tree is a big one, it can leave behind quite a sizable gap in the canopy, which allows more sunlight to reach the forest floor and stimulate pioneer growth. It is often recognized that natural treefalls play a critical role in the cycles of forest growth and regrowth. They cause much alarm. Schebesta describes the scene one night in 1924/1925 in a Jahai camp: "A crash, as though the earth beneath our feet was bursting asunder, made me jump up. Terrified, I looked out into the blackness of the night. At that moment the camp fires burst into bright flames, as the smouldering embers were stirred into life."[8] Kirk Endicott writes: "The heavy downpour of rain can loosen the roots of the shallow-rooted rainforest trees, and the sudden gusts of wind can blow down large branches and can topple even the largest of trees. . . . With each gust of wind, the trees groan and crack. The air is filled with shouts of warning, and families rush from one shelter to another."[9]

Forest clearings caused by treefalls are marked in the language: *pərban* seems to be the generic category for such clearings, while *tupan* are those clearings where the fallen timbers are rather large. There are terms that record gradations of tree-decay, this being one way of determining salience. *Rənbak* is any newly fallen tree that has not begun to rot; *təras* a dead, long-rotted tree; and *rel*, a still-standing but dying tree. These are common features in the forest: used cognitively as landmarks or signposts and practically for such purposes as river-crossing (see appendix B). Once a tree has begun its slow decline (entered the *rel* category) or fallen (become a *rənbak*) it eventually will become a *təras*. If it falls over a river, it is a convenient bridge and its salience as landmark will increase until such time as it collapses and people look for another another convenient crossing-point.

In this lowland ecosystem, perhaps only swift-flowing runnels are not so common. Otherwise, all types of waterflow from narrow rivulets to broad rivers occur. Given the relief carved by this extensive network of watercourses, movement through the landscape is punctuated by a series of lurches, up and down, up and down, from land down to water and up again (see, for example, the iteration of movement verbs in appendix B). In evaluating one route over another, the different parts and characteristics of a watercourse are extremely salient. For example, the *mɔs tɔm* (upper reaches of rivers) are rejected as camping grounds because they are associated with disease spread; they are also food-poor. The most common pattern in movement is to travel through the headwaters of minor drainage systems. Most of the tributaries and feeder streams are not navigable by boat. But these, rather than the major rivers, are the true waterways. Major rivers, like the Təmiliŋ and Kəciw, are positioning anchors, meaning that travelers at all times know where they are in relation to them and which tributary system they are in. During the

Plate 5. Mounting a *rənbak* (newly fallen tree). Photographed by an anonymous Batek youth, 1996.

course of a season's movement, a camp group might end up camping on the banks of such rivers.

Using kinship metaphors,[10] the Batek seem to classify river systems hierarchically. They will say that there are only two classes of rivers: *tɔm naʔ* (lit. mother river) and *tɔm wɔŋ* (lit. child river: i.e., tributary). To label an affluent (tributary of a tributary), they might add the diminutive *ʔawãʔ* (whose primary meaning is "child") to *tɔm wɔŋ*, thus forming the subcategory of "little child river." The "mother river" of one tributary system may be a "child river" of a larger tributary which is in turn the "child" of a yet larger "mother river." Accordingly, all the rivers in a drainage system will be nested together in reciprocal child-parent relationships. Topographic contiguity is assessed by knowing where rivers "head" towards (*kiy-kuy*, derived from *kuy* [head]) or meet (*pibus*) each other. In movement, the trick is to know where each river joins up with its source.

Organizationally, the opposing points of a river are its *mɔhʔɔŋ* (source) and *was* (confluence; mouth). Perhaps due, again, to the lowland nature of Batek adaptation, *mɔhʔɔŋ* is not so frequently encountered. The *was* (Malay *kuala*) is most salient, being convenient points of camping and (for external folks in the past) fort construction and village settlements—note for example how many

Malaysian place-names have *kuala* as the forename (Kuala Lumpur, Kuala Lipis, Kuala Tembeling, etc.). At the point of the *was*, there will be a fork (*jəniŋwaŋ*). Moving in headwaters areas, however, the Batek are always crossing rivers wherever there is a convenient point, and not only at the *was*.

Most of the time, the incipient forest traveler will be totally bewildered. The rivers are rarely going to be straight (*bətow*); more often, they are crooked (*kəliŋwəŋ*) or meandering round and round (*bumutlit*). These adjectives are also used to describe walking trails, of which more shortly. It will seem that the river is flowing round and round (*kunah*). It is likely, for instance, that quite a few of the *ʔalor* (stream channels) Lus mentions in appendix B are different parts of the same watercourse. As the rivers snake around the landscape, they will exhibit features like *təloʔ bəw* (big bend in the river); *təmbuah* (deep waters); *tɔm liwin* (waters turning slowly round and round); *gul* (slow-moving waters); and so on. There will be piled up driftwood and leaf litter (*rəmram*) here and there and as people move through the landscape they might point out the many places on the riverside (*təbiŋ tɔm*) where they once set up a fishing rod (*rənɛm*).

Water level goes up and down, of course, depending on the rainfall. In the Peninsula, rainfall patterns rather than temperature changes are the cause of seasonality. According to taʔKadɔy, the rainy season (*masaʔ banyir*) is marked by two terms: *can banyir* (lit. foot of the flood) is the onset and *ləspəs ʔuʔ ral* is the end of the season. The effects on river flows are marked with a couple of terms: *tɔm zĩl* (flood season river, when it floods its banks) and *tɔm kəmaraw* (dry season river). Seasonal patterns aside, the rivers could be described thus: big or swollen (*bəw*), clear (*cəŋrəŋ*), muddy (*pərkac*), and so on. During the flood season (around the end and start of the calendar year), groups might disperse into even smaller groups (population around twenty-five) and choose to camp out in *teʔ laŋeh*, away from the worst effects of flooding in the *həp ləy*. At this point, hunting is said to be good since the game animals are less active and more easily shot ("they just sit there," as I was told). Movement out being cumbersome, the groups might live off forest resources entirely until water levels go down. During the dry season, features like *hə̃nril* begin to show up: this is the strata of a riverbank wall that is exposed when the water level goes down.

The terrain is also represented in the lexicon. In these rolling hills, *papar* (slope; small ridge) is a common feature. Where slopes meet, the ridge top or ridge crest may be in the form of a plateau, a *hnadaŋ*; this is also considered the natural boundary of a watershed. *Hnadaŋ* can be high or low. They are in the *mɔs* (upper reaches) and therefore part of the *teʔ laŋeh* landscape. I have camped in a *hnadaŋ* just once in all these years and the reason is clear: the water source is not usually close by (any camp leader proposing to take the group to a *hnadaŋ* will have to make an unassailable case). However, many *hnadaŋ* can be pleasant to walk on once one has made the full ascent. At least one *hnadaŋ* in the Ralat area, where Tanyoŋ Baŋlas and Tanyoŋ Rənam meet, even has a song to celebrate it. (*Tanyoŋ* refers to any land type that juts out like a peninsula, such as the capes of rivers, promontories, or the ridge crests. *Rənam* are fragrant leaves collected for bodily ornamentation. *Baŋlas* may be a plant name.)

Higher than ridges are the mountains. There is some controversy on this point. TaʔKadɔy recognizes four kinds of hills and mountains: in order of altitude, *cəbaʔ* (hillock or ridge-crest), *bukit* (hill; a Malay loan), *gunoŋ* (mountain; a Malay loan), and *bənəm* (very tall mountain). This was refuted by ʔeyTəltil (who was, however, unable to explain how each of these four terms is distinctive). Semantically one could even say that a *bukit* is only the Malay word for *cəbaʔ* and a *gunoŋ* the Malay word for *bənəm* (and vice versa). A "splitter" like taʔKadɔy, who has a wealth of landscape terms and knowledge to share, can recognize the differences, which may be fine indeed; a "lumper" like ʔeyTəltil, who in recent years has been spending less and less time actually traveling in the forest, sees no need to.

Bənəm is the classic Mon-Khmer word for "mountain." It is, for example, the origin of the name of Gunung Benom (in Chewong and Jah Hut territories), which therefore means "Mount Mountain"! I suspect that *bənəm* represents a truly high mountain that the Batek would never, unless guiding tourists, ascend in the ordinary course of life. It therefore has mystical associations, like so many mountains. For travelers, distinguishing one kind of mountain over another may not be as important as the practical issue of locating (other than the lowlying hills and ridges) the passes or gaps between hills and mountains, the *wɛ̃c*.

In the course of a path, it takes on different qualities. As the terminology shows, path qualities are dependent on concepts of movement, a point that Burenhult has also revealed for the Jahai.[11] Normally, any walking path is a *halbəw* or *harbəw* (see below for more discussion). When it slopes upwards, it becomes a *cəniwəh*, which is derived from *cwəh* (to ascend). Correspondingly, on the downward slope, the path becomes a *pənisar*, derived from *sar* (to descend). Then, if the path winds round and round like a switchback, zigzagging around the contours of the land, it becomes a *pənitər*, derived from *tətər*, which means to walk on such a path. Finally, if the trail followed is a short detour from the main path, it is a *pəniwəh*, derived from *tiwəh* (to walk around obstructions). As the path approaches a branching point, it may be called a *halbəw niwaŋ*. *Niwaŋ*, whose meaning is not clear, is almost certainly the root word for *jəniŋwaŋ* which refers to any kind of branching point, be it of overland trails, rivers, tree branches, fingers and toes, and so on. Possibly *niwaŋ* is the verb—to fork—and *jəniŋwaŋ* the noun—the fork itself. There are many other terms that describe types, parts, and conditions of paths: *bəldɛ̃l* (broad path*)*, *bəplit* (longer route, i.e., not a shortcut), *gitʔac* (ground soft and muddy from rain), *həŋʔuʔ* (overgrown), *pərəŋdəŋ* (broad path), *paw halbəw* (trailside or verge), *tə-bɔ̃ŋ* (start of a path), *təbeŋ* (rise in the path), *trichõc* (very narrow path), and so on.

Landscape perception is of course intimately tied to its biological characteristics. Mention must be made of the other inhabitants of the forest, the plants and animals. The Batek have a wealth of observations about these species, their morphology, functions, relationships, ecology, and growth habits. Following the theme of this chapter, I provide here a brief introduction to the classifications and taxonomy.

As Kirk Endicott observed, a key criteria of classification is whether a species is edible or not (see chapter 4 on the origin of animals). In line with the concern over gastronomy, how a species is eaten is an important taxonomic criterion: so fruits (*plo?*) would be anything that is *lɔt*; meat (*sec*) whatever is *rɛɲ*; carbohydrates (*bab*) whatever is *ci?*; and vegetables, palm piths and cabbages (*ta?a?*) whatever is *hãw*. Which encourages verbal economy: "*Yɛ? yik kan lɔt* [I don't want to *lɔt*]" is understood to mean "I don't want to eat fruits." Correspondingly, as Endicott pointed out, "if a person says he will *rɛɲ pacɛw* (eat water monitor), it is clear that *pacɛw* must be a kind of animal."[12]

The traditional starch staple was *takop*, the generic term for wild yams taken from the name of one common species, *Dioscorea orbiculata*: "the expression *bay takop* (dig *takop*) is commonly used to indicate the process of collecting wild tubers in general and, even more broadly, of living off wild as opposed to cultivated plants."[13] The Batek's favorite foods are, however, the seasonal forest fruits. The generic name for these is *tahun* (from the Malay word for "year") or *kɔbi? tahun*, which is a subset of *plo?* (edible fruits) that are in turn a type of *kɔbi?* (fruits). Though vegetative foods probably provide the bulk of the traditional diet, the Batek's perception does seem biased towards animals.

Animals have an honored place in their imagination; the Batek are, like many hunters,[14] enraptured by animals, first-rate observers of and intellectually stimulated by animal appearances, habits, and behavior, a point I return to in the following chapters. Batek interest in animals goes beyond satisfying gastronomical needs although that may indeed be one basis for their knowledge. Kirk Endicott remarks:

why these terms? what vision of people?

> Batek ideas about most of the animal species they eat are basically rational and well-founded. They have detailed knowledge of the anatomy of these animals, although, as in the case of humans, they do not correctly understand the functions of some of the internal organs. Their knowledge of the living habits of the different species is astonishingly extensive and accurate.[15]

The category *?ay* encompasses mammals, reptiles, amphibians, fish, birds, invertebrates, and insects—in fact anything nonhuman that moves, including lice and ticks. Another word for animals is *mɔnantaŋ*, a loan of Malay *binatang*; this is not used as spontaneously as *?ay*. Game animals may be specified as *?ay hi?* (our animals). There is no comparable word equivalent to "plants," which are identified through species names and grouped into a few genera: yams, fruits, trees (*kayu?*), bamboos (*buloh*), flowers (*boŋa?*), and all woody and nonwoody vines, including rattans (*?awey*). Within the kingdom of animals, the generic categories are: birds (*kawaw*), fish (*?ikan*), snakes (*?olar*), and, it appears, butterflies (*tawãk*). As revealed by the tendency to borrow Malay words for generic terms (see notes for these words in the glossary), there seems little need to talk about species as members of groups; most are identified through monomials (single-word names). Crosscutting the genera, three main animal divisions seem to be recognized: covertly, big and small creatures, and explicitly, the predators (*bẽc*). This last group is symbolized by the much-feared tiger (*yah*), the archtypal predator (it will make a full-scale appearance on pp. 111–17).

The largest body of animal names in my collection is for birds; the names were often given spontaneously to me, thus suggesting that birds are often in the Batek's minds. At least two species, the sandpiper *kədidiʔ* (probably *Actitis hypoleucos*) and the spiderhunter *sətsɛt* (*Arachnothera longirostra*) have important roles in different parts of the genesis cycle; these are considered *halaʔ* birds. Kirk Endicott also heard that the knowledge to distinguish edible from inedible fruits (and to process those which are unpalatable or poisonous) was revealed by bird-form *halaʔ*.[16] Using standard classificatory terms for special animals, there are several recognized bird categories: *kawaw ʔasal* (original birds); *kawaw halaʔ* (superhuman birds); *kawaw tahun* (fruit season bird); *kawaw hantuʔ* (ghost birds); and *ləʔ* (indicators).

Much of Batek ritual is about keeping categories distinct: that between humans and animals and, with respect to habitat, those of forest, non-forest, land, water, sky, underground, and treetop.[17] Most plants and animals can be fitted, if roughly, into one of these habitats. The most troublesome—most salient—creatures are those that can't be boxed in one category: they are in-between creatures. Perhaps they spend the most time in one niche and feed off another, or habitually inhabit both land and water or treetop and land. They are ambiguous and, following Mary Douglas, have "polluting" and possibly ritually dangerous status.[18] Also salient in the same way are animals that behave in ways not befitting their like. For example, the flying lemur and the flying foxes are "non-birds" that fly and these are usually mentioned together in the same breath.

Further delineation of the ambiguous and possibly polluting status of such animals comes from foraging beliefs. In one, it is said that a hunt will be *malaŋ*, crippled by ill-luck, if the hunter encounters a variety of animals like flying lemur, land monitor, snake, and mud-turtle. What links all these animals is that they inhabit multiple niches (both sky and treetop or both land and water). By far the most ritually salient animals are the *bawac* (pig-tailed macaque; *Macaca nemestrinus*) and the *jəlew* (long-tailed macaque; *M. fascicularis*). *Bawac* travels widely in troops and feeds both on ground and in the trees. *Jəlew* is a renowned pest to farmers and fishermen throughout Southeast Asia and is easily acculturated to man-made landscapes; it is the monkey most often seen (and fed) in Malaysia's parks and temples. These two monkeys have prohibition myths that warn people not to mock them, either in words or deed, or there will be thunderstorms. *ʔeyJudiy* even considers that the myths put these two animals in their own class.[19]

Some wildlife are too human-like to be mimicked. These macaques, especially the *bawac*, belong in this category. *Bawac* has a singular characteristic that defines ambiguity: among the animals, it has the most human-like facial appearance. It looks like a human trapped in the body of an animal. Its prohibition myth (which follows a motif common among Orang Asli and in Borneo) warns against treating it like a human.[20] It remains one of the least hunted of all treetop game. Especially during rainstorm spells when the cosmos seems particularly unstable, hunters might resist capturing the *bawac* so that unwary children have no chance to make insulting remarks about it and therefore anger the thunder-god.

ÁH¹⁰ ⱦ 8A ₴ JⱧ
EB

5ᵘ

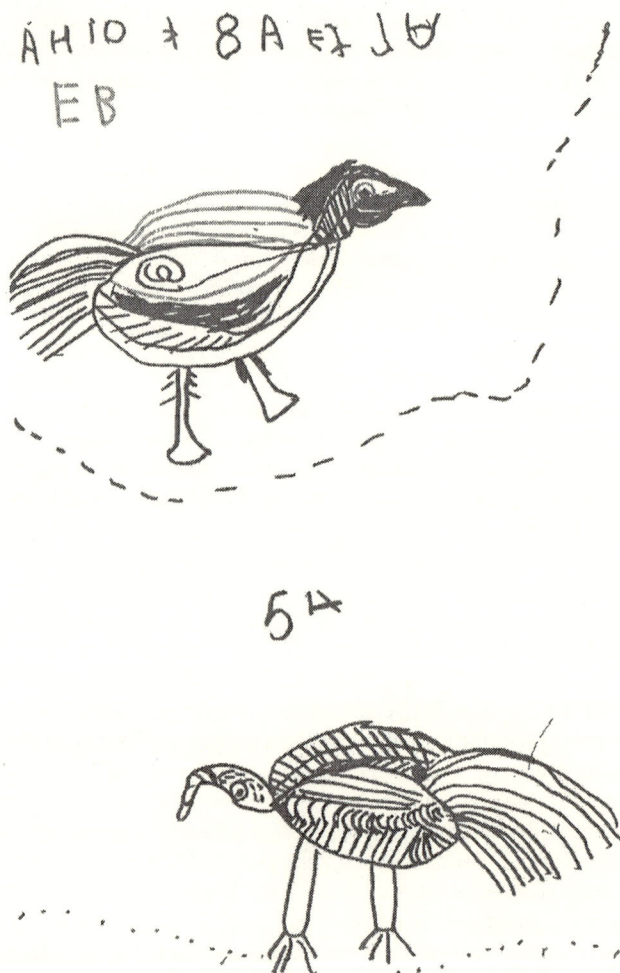

Figure 3.1. Animals have an honored place in Batek imagination, as revealed by many
adult and child sketches. This depiction of pheasants and their trails is from the hand
of Kadɔy (a nine-year-old boy), 1996.

One basis for the form of Batek environmental knowledge is related to their
style of "broad spectrum foraging." In broad spectrum foraging, almost everything
is potentially useful. Not everything has been identified with name and use but
nothing should be discounted in case of future values. Who knows what new
economic context might turn up and something that's practically useless today
becomes useful tomorrow? This kind of perception, readiness to switch course
and draw on different kinds of resource knowledge for different purposes,
discourages specialization in a few narrow domains. The Batek are collectors and

generalists *par excellence*. Some groups of plants and animals have special qualities and recognized salience. But the vast majority are not subjected to specialist treatment. Philosophically, the Batek seem unwilling *not* to know about anything that comes across their way. To paraphrase Kirk Endicott,[21] the Batek are foragers of information and store an inexhaustible body of details about the forest and its changing conditions.

Pathways

As the foregoing suggests, the paths are critical to Batek uses and concepts of the forest. They merit separate discussion. The words *halbəw* or *harbəw* have a dual meaning, referring not only to the walking path itself (the track or trail), but the notional "way" or "route." "Pathway" captures the fact that *halbəw* is as much concept as material reality.

The "pathway" is a most serviceable metaphor in any language—and quite a few religious orders—and so too for the Batek. The importance of pathways to hunter-gatherers has received more attention in recent years. The best known examples are from Aboriginal Australia. Material from there shows that conventional land area boundaries make little sense to hunter-gatherers. Where you journey, there is your country. And your country is marked by the passageways of your trails. Wagner has captured this point well: "the life course of a people, the totality of their ways, conventions, and conventionally encountered situations, is the sum of its 'tracks,' the trails over its country along which experience is measured out" (interpreting Nancy Munn's material on the Walbiri of Australia).[22]

As mentioned, environmental knowledge is organized around the pathways and the places linked up by them. The question is how the knowledge is obtained in the first place. One way to find out: observe the Batek walk along a trail. A walk is an event in itself, an active process of perceiving, discriminating, discussing, and teaching. Contrary to the image of fleet-footed forest dwellers that is so pervasive in the ethnographic literature, often the Batek sacrifice speed for the sake of moving leisurely. Sometimes they may adjust their speed of walking to the level of the least competent member of the group; in many cases during my fieldwork, that person would have been the note-taking, foot-fumbling anthropologist in their midst! The order of the travelers to some degree mimics the order of competence, with children going in front and the men providing the rearguard, thus enabling them to keep a watchful eye over the others (tiny children are, of course, carried on their parents' bodies).

Sometimes the talk is muted while at other times it is raucous. Even as they move, they are discussing other, non-immediate issues; by the time an excursion is over, some kind of a resolution or decision might have been made. If those in front are unsure of the directions, a stream of instructions is called from the back. Children are discouraged from talking too loudly and warned that the tiger or other predators will hear. Ears are trained to hear noises from afar; if the noise is significant enough, this is discussed as well. While they walk, they keep their eyes directed to the front. As ya?Kaw declared, the way she keeps herself from getting lost is to look right ahead, follow the tracks pointing to the main trail, and not be distracted by side-paths going here and there.

When they spot any material of interest (like game or plant foods), they either discuss it casually or ponder aloud whether they should procure it. Sometimes the movement is halted when members try to persuade one of the group to extend that physical effort. When someone spots something at the path-edge, such as the trailing vine of a yam, s/he will inspect it carefully and the whole pack might halt its movement to discuss this new finding. As new opportunities arise for culling things out of the forest, the order of travelers will change and some in the group might linger behind to follow through while the others precede them to an agreed place of collection or extraction. If young children are in the group, the parents might point out things of interest to them and if the children demand to have something brought to them, the movement will be halted for this purpose.

What I'm indicating here is the range of experiences and sensations afforded by pathways. Walking on a trail, the Batek are actively monitoring it. They are looking at changes in the vegetation, spotting animal spoors and traces, indicators of change, swapping information—and affecting the characteristics of the trail by the very act of using it. Weiner's words are entirely *apropos* here: "Journeying through Hegeso territory on foot is never a matter of merely getting from one point to another."[23] Similarly, a pathway is not just a line between departure and arrival. It is a route to knowledge.

Repetitive, habitual movement over a network of well-defined trails is the method of acquiring familiarity. I asked ta?Jamal how he had learnt to travel from one place to another. He said the lessons had come from his father. First his father permitted him to *jok* on his own or with companions for short distances only. As he grew older, so did his travels lengthen. In the beginning, he was afraid to travel independently. But as he grew older and more familiar with the environment, he said, he began to feel less afraid. His father let him travel farther and farther. Eventually, he was no longer afraid to travel on his own. Maturity was thus equated with the capacity to travel far. As ta?Jamal's recollections show, learning to attend to the important details of the environment is not possible unless it is done consistently and continuously. Understanding is a lifelong process, begun ideally from the first moments of life.

Descola has described this for the Achuar of Ecuador: "a system of esoteric markers must be called upon which requires an intimate knowledge of the salient elements of the micro-region in question: a peccary wallow, a salt lick regularly visited by animals, a deposit of pottery clay, an exceptionally large tree . . . or localized stands of [specific plant communities]."[24] The difference between the Achuar and the Batek is that the former are horticulturists, for whom daily life is centered on the house and its surrounding gardens. Thus, as Descola goes on to say, "as the distance from home increases, the forest gradually becomes a *terra incognita*, devoid of familiar markers."[25] This happens less for the Batek because as long as there are paths and people to move along them, then there is no true wilderness.

A social and productive perspective on pathways, showing people walking to and from camp and (top left corner) a hunter confronting his prey. The path follows the course of the river. Anonymous artist, 1996.

A stylized model of a pathway from Yip (a twelve-year-old boy), 1996.

Figure 3.2. Two Batek depictions of pathways.

The classification of pathways is one way to represent land use. If pathways are spatial expressions of movement, their layouts will betray schedules and habits. Further, there should be ecological and cognitive effects to activity patterns. Foraging activities are most focused on trailsides: from the trail, you may spot useful materials. People don't go randomly searching for things without a certain plan or direction in mind. They don't just take any trail they please. The well-trod ones are used most commonly—up to a point. When gathering or collecting in the forest, the tendency is to comb through looking for vines and vegetables, thus necessitating leaving the trails and opening up alternate passageways. The general strategy, which seems to be most time-effective and safest, is to travel to the farthest possible point on the main trail and then begin the search-and-harvest backwards in the direction of the campsite. The Batek say that, to prevent getting lost (*cirloŋ*), they should at all times orientate themselves in relation to the river and to the campsite.

With further action and as topography changes (for example, a path is revealed to be too slippery after rainfall), then new pathways are opened up. "Opening up" may be an extravagant way to describe the practice. The Batek do not cut openings unless the way is impossibly blocked. Typically, they slash a few vines and branches and the vegetation will "close up" behind them. Following on their heels, it can be hard for an outsider to see where exactly they have pushed through the vegetation unless the passageways already have a well-worn aspect. Path cutting, as with the clearing of lean-to space (see below), is done ever so casually. Often the sharp points of cut saplings stick up at shin level. These sticks provide a natural grip for little children as they learn to manoeuvre their way around the forest.

There are three overlapping sets of pathways. Those within forest bounds the Batek sometimes signify as *halbəw Batek* (Batek pathways). Since they spend most of their time in the *həp ləy*, we should expect to find the densest network of pathways there. As we saw earlier, the *teʔ tərap* contains the most overgrown trails while the *teʔ lapeh* have the most deadends or trailheads. Among those that have become "public" or fragmented, we would include the logging and access roads that are cut through the Batek trail system. In the national park, the Batek helped to open the trail system for tourism purposes and therefore the tourist trails tend to mix with and overlap the Batek trails.

A third set of pathways are those that criss-cross in the vicinity of the forest camps. Aside from the *jaman tɔm* (path to the river), there are those leading from one lean-to to another within the bounds of the encampment, and those leading out to the forest. The first kind of pathway comes into being over the course of daily projects and visiting. The second might be historical, a track that is part of the broader system connecting one encampment with another. The pathways that lead to a campsite are the same ones leading towards foraging zones. When lean-tos are moved, new passageways are broken in; tangles spring up in different places around the campsite.

As the environment changes, so too the routes and the Batek's perceptual bounds. This is shown in a development of terms to describe newer categories of paths: *halbəw tah* (asphalt or surfaced road), *halbəw wiŋ* (logging road), *halbəw pəlancoŋ* (tourist trail), and *halbəw banar* (town road). In these binomials, the secondary words are of foreign origin: *tah* from English *tar[red]*; *pəlancoŋ* and *banar* from Malay *pelancong* and *bandar* (the mysterious "*wiŋ*" has not been defined). Following this logic, a road is just a different type of pathway.

Practically, knowledge of the pathways enables individuals and groups to plan their own movements and keep track of others'. Hence it is not critical to know *precisely* where someone is living right now so long as one knows where he or she generally is in relation to the pathways (particularly the rivers in that location). Knowledge of space and location further enables people to monitor the ongoing state of the ecology and of each other. Because of the high degree of flux and intercamp visitation, new findings and discoveries about the social and ecological worlds are often shared across vast distances in the form of news and gossip exchange. Additionally, news and gossip serve not just to monitor the resource base, but also to maintain ongoing links between people who are far apart and hence to give everybody a sense of community and belonging.

Camps and Villages, Lean-tos and Houses

The previous sections have introduced the setting with an expansive view of the landscape: from the concepts of the "world" and "land," we entered the forest and examined topographical and ecological terms and features, ending up with movement and the use of pathways. Of the three important cultural symbols mentioned earlier (p. 54), what's left are the camps. These represent the world of people and are interconnected by the pathways. We have encountered some key terms already; it is time to probe how their various meanings relate to each other, particularly in their stereotyped and idealized forms.

There is a set of forest characteristics that is understood to exist by virtue of what it contrasts with. *Həp* may be contrasted to *dəŋ* (Malay house and village; a cultural, social, and vegetational contrast) or *tɔm* (river; a topographic contrast), and, at the micro-level, to *hayã?* (lean-to and camp; a spatial contrast). In other words, when the Batek think about the *həp*, they may be thinking how it is different from the world outside as represented by the village (the basis for their identity), how it is different from the watercourses, and/or how the forest is just beyond the bounds of the lean-to. These contrastive pairs are strongly highlighted in language and belief.

One central contrast is between *həp* and *dəŋ*. The Batek cannot distance themselves from the forest or view it as peripheral to society; to produce an objectified vision of the forest, they evaluate it through the mirror of the *dəŋ*. The word *dəŋ* has an impressive pedigree. It goes all the way back to Proto Mon-Khmer (i.e., perhaps five to six thousand years old) and the cognates remain widespread among Mon-Khmer speakers; the original meaning is "house on stilts" (in old Mon, the meaning was extended from "house" to "village" to the political and notional categories of "nation").[26] For the Batek, the central meanings are "village" and "house." The etymology suggests that "house" was the primary meaning that, following the standard pattern, then came to encompass the settlement pattern associated with permanent houses, including, today, towns and cities. Following the regional pattern in Southeast Asia, the Batek equate forest with upriver (*bah-kəntəʔ*) and village with downriver (*bah-kiyɔm*, which also means "underneath" or "below"). Thus to *sar bah-dəŋ* (go down to village/town) is contrasted to *cwəh bah-həp* (go up to the forest). To the Batek, *dəŋ* are centrally identified with the Malay world. And the Malay world, as will be discussed below, is viewed in predatorial terms.

The architectural forms in the villages, the houses, are sharply contrasted to the lean-tos in camp. Houses and lean-tos, in short, are ethnic markers and there is much evidence that the Batek have clear ideas what types of dwelling are appropriate for different evironments. This contrast is morally charged with ideas of value as shown in quite a few Batek folktales that recount the unlucky ends of forest peoples who visit houses. Consider the oft-told tale of Paʔ ʔaŋkol, a clever shaman. The story begins with his sisters who wandered too far downriver:

> The younger of the sisters was afraid, "We are near the home of yaʔKədat[27]. Let's go home." The other sister didn't believe her. She wanted to go to a *gob*'s house to eat betel nut. They came across a house. Thinking it was a *gob*'s, they went in. However, this was yaʔKədat's house. At first, she welcomed the girls and gave them betel nut. After they had taken this, however, she took a machete and killed them both. Then she ate them.[28]

Such stories enforce the negative impression of houses. In history, slave-raiders posed the greatest threat. Cannibals may in fact symbolize slave-raiders, which would give such stories extra significance: they are cautionary lessons warning against misplaced trust.[29] Given this history, *dəŋ* has negative associations—it is a place that forest peoples venture into at grave peril, for they might never return from there.

Another contrast is revealed in the oft-stated distinction between *cip bah-həp* (go to the forest) and *wek bah-hayãʔ* (return to lean-to/camp). The word *hayãʔ* is also of Proto Mon-Khmer origins and has a widespread distribution today. Its etymon also means "house." For example, it is cognate with the Vietnamese word for "house," *nhã'*. Its principal meaning is certainly "lean-to" (or perhaps, to be etymologically correct, "house on the ground"). The *hayãʔ* is highly valued. We could not overestimate its significance. A true forest-dweller is recognized to be a lean-to dweller. When the Batek tell stories of their travels, they almost always refer to the initial construction of the lean-to and the setting up of the camp. All references to the habitations of non-visible beings use the word *hayãʔ*. The fruit deities are believed to have their own lean-to in the cosmological center. There is no separate word for "camp," which would show that the camp is considered just a larger extension of the lean-to (as the village is just a larger extension of the house). The primary opposition, then, is between the dwelling, the home, and the broader forest beyond.

Plate 6. Kechau campsite, August 1995.

Rivers (the *tɔm*) are territorial and orientation markers; practically, as we have seen, they are convenient for reckoning direction and position. In the sense that they create discontinuities in the land, they are of the forest and yet also apart from it. Rivers give identity to places and, by extension, the people who come from there. The names of encampments are typically those of the rivers or streams that flow by them. Topographically, forestland ends where the water flow begins. By default rivers are "the end of land" or "an interruption to the land" and all rivers are salient by the fact that they are not-*teʔ* or not-*hɔp*. To recall a discussion from chapter 2 (pp. 37–38), rivers can have negative associations. If the world is an island, rivers are the arteries connecting the land with the shadowy and watery underground. As such, rivers have their threatening quality: when they are swollen, they may wash the land away or soften it so much that it collapses. There is a wealth of taboos (especially those classified in the *lawac* category[30]) that highlights the special distinction between land and water. One, for example, forbids letting certain kinds of blood, like the blood of ritually salient animals, flow into the rivers or bathing in the river after eating certain kinds of meat.

This discussion shows that the Batek carefully categorize what may be found in or are associated with the forest. Only *dɔŋ* definitively symbolizes the outside world. From the perpective of the language, forest is contrasted to the world "out-there" (village) and the home "in-here" (lean-to; camp). And since the "in-here" is always *of the forest*, never outside it, the forest can never truly be externalized—so long as the perceiver remains in the forest. Because there is the world of the village to contrast the forest with, the forest can always be conceived as an abstraction: i.e., it is "X" because it is "not Y." As Dentan warns of similar plays of opposites in Semai thought, "it is important not to think of [these concepts] only as opposites. They are opposites, but by virtue of being joined as opposites, each implies the other."[31] They are equivalences, and the very fact of the contrast shows how much the village is part of the definition of forest. This is not to claim that the forest could not exist without the village but that so long as there is a concept of the village, so is there a mirror upon which the image of the forest is refracted.

Degradation of the Ideal

What are the environmental relations implied by these terms? The study of culture/ nature relations has spawned an encyclopaedic literature, some of it verging on caricature. Not surprisingly, the Batek do not have local words for either "culture" or "nature." But, following their tendency to compare and contrast, it is useful to probe their version of appropriate people-forest relations, if only to contribute to the general literature and, for the purpose of this book, to highlight why some environmental processes provoke so much worry. I focus here on the contrasts between "house/village" and "lean-to/camp" and offer my own interpretations of how they symbolize contrasting conceptions.

The *dəŋ* and the *hayã?* contrast with one another and more broadly each is also the reverse image of the forest. In its core meaning, *dəŋ* is a neutral word. Any physical structure that resembles a house or house-on-stilts is a *dəŋ*. Within the broader framework of meaning, however, the cultural references are negative. As "village," what is distinctive is the type of vegetational cover: villages are agrarian spaces cleared from forestland and therefore, by definition, opened-up and transformed areas. Camps do not have the luxury of escaping the forest. Spatially and ecologically, they are part of the forest: built largely from forest materials, located within forest bounds. They represent the world of people; they are in the "foreground" of perception.

Social practices are a relevant clue to perceptions. The traditional Malay dwelling, whether in its pioneering or established form, was not spatially distant from the forest. But the Malay tendency was to "take" useful materials from the forest, as seen in a geographical survey of Malay communities in the 1950s: "Although the people of these communities frequently employ plants that have been derived from the forest in the past, these are today cultivated in the garden or near the *kampong*."[32] In Whitmore's list of cultivated plants from Kampong Melor in 1971, we find twelve species of fruit trees that are also found "wild" in the forest, and several other cultivated varieties that have "ancestors" in the adjacent forest.[33] I would say that in horticulture, Malay villagers reproduced the forest in village bounds; i.e., they created an image of the forest in the village.

Camps, on the other hand, have impermanent structures. Field farming or systematic tree-crop planting are not done. If we take a cross-section of a campsite, the landscape ranges faintly (on a gradient from "culture" to "nature") from the lean-to on one end to the intermediate category of the camp, to the broader forest beyond. What constitutes "forest" is overtly anything outside the lean-to but covertly is also a matter of perspective. Sleeping without lean-to cover, though one may remain within the bounds of the camp, is reckoned as *tek hat həp* (sleeping in the forest). But if, say, one was already at camp, standing outside the lean-to, to leave camp for the forest is still to *cip bah-həp* (go to the forest) though one is, technically, *already* in the *həp*. There is no true difference between camp and forest; what matters is the *direction* of movement; the daily round of toing-and-froing is what makes the forest different from the camp. There is no need to reproduce an image of the forest in the camp; rather, there is a tendency to reproduce the camp in the image of the forest.

Spatial distinctions are blurred in a camp. The boundary between lean-to and forest is a moving one; it is truly the fact of people dwelling in the lean-to that gives the boundary any logic. Lean-tos are somewhat open to the elements. Karen Endicott describes: "These shelters are open on three sides, thus exposing the activities of camp members to the full view of everyone else in camp."[34] If side-walls or frontal overhangs are not built to keep out the rain, it is possible to see the forest almost every which way one turns (a lean-to "wall" is a vertical arrangement of thatch layers, *hapəy*). Moreover, a lean-to's construction never seems to reach finality. Alterations and improvements may be made to the basic framework and thatch covering. Only when it's time to move to another camp can a lean-to builder put the stamp of completion on her work. Social relations may be an additional factor. One (to me) rather amusing example concerned the lean-to building adventures of two sisters aged about ten. They got into all sorts of squabbles. When they couldn't stand each other, they'd uproot their lean-tos to face opposing directions (back to back—a bit like slamming the door in another's face); when peace negotiations ruled, they lined up the lean-tos in adjacent or facing directions. In other instances, lean-tos may be moved around the campsite over the course of residence (for example, after a rainfall, when water seeps in; see Plate 7). As I see it, the camp always threatens to emerge but never quite makes it. Small wonder, then, that there is no separate word to call it by. Kirk Endicott made a similar observation when he declared that "the Batek, unlike most forest-dwelling agricultural peoples, do not attempt to carve out an island of culture in the sea of nature."[35]

What is common to both village and camp is the need to make space for people. In the village, space clearing is taken perhaps to its extreme. A common view of Malay villages is how neatly kept and physically ordered they are.[36] The forest is *always* in the background. As for Batek camps, the image that I use to describe the process of camp formation is one of "folding back" the vegetation. Young trees are cut, vines and lianas are severed from the rootstock; enough room (spatially and vertically) is opened up. The Batek make few attempts to empty the lean-to surround of vegetation. Hence the image of camp establishment, as one after another household unit sets up its lean-to, is one of clearing space in the forest but not of *cleaning* it, quite unlike village establishment.

But does this suggest that the Batek do not order their environment? Their animal classifications recognize a clear distinction between humans and

Plate 7. Relocating a lean-to, September 1995. A heavy rainfall the night before has forced this family to site their lean-to on higher ground. While the parents pause in their work, the little girls are bouncing on the newly laid floor.

nonhumans. This distinction must be maintained at the risk of inviting cosmic calamity. Following this symbolic order, the lean-to represents people and the forest the nonhuman world. As such, the forest can be regarded as "background" that threatens to subsume the foreground of daily life. For to forget the distinction, mocking the boundary between human and nonhuman, is to invite the wrath of the spiritual forces and, ultimately, land collapse. This is the central conundrum: that the world of the camp and the world of the forest are integrated within the same social-spatial frame but must be kept at a distance from one another. This seems to be a classic way of relating to "nature"; to be aware of and yet not distanced from its capacity for threat.

The villages and camps represent similar yet different ways of accommodating to the luxuriant vegetation of the forest. So long as population densities are low, and there is no capital investment in "developing" forestlands, both can co-exist. However, that has not been the historical trajectory and, as I pointed out in chapter 1 (p. 8), the village is also becoming absorbed into the city. Through the tumult of history, lowland settlement has tended to start off along the coastlines and at estuarines (the Malay *kuala*; the Batek *was*), moving gradually landwards into the hilly interiors. Today the highlands account for most of the remaining forests in the Peninsula. The more extended the village (and its counterpart, the town), the smaller the forest. The village is predatorial. The smaller the forest the less room there is for the camps. And the forest is subject to scrutiny by the forest peoples more intensely.

In sum, though the concepts of lean-to, house, camp, and village represent ideal-type images, they are really dialectical concepts whose meanings can change through time. In Dove's study of the *jangal* concept in South Asia (the etymon of "jungle"), he proposes a similar evolution of meanings: that the *jangal* comes to seem more "wild" the farther society retreats from it.[37] We'll find a similar theme among the Malays (see p. 104). As Kato has argued about the traditional belief system, "the forest was not simply a collection of huge trees. It was inhabited by good as well as evil spirits; it was feared and awed; it was considered to be a source of potent power." But with the colonial-inspired burning of forest tracts to open up land for plantation estates, and as Malays were urged to put aside their "traditional" ways, so did they start to lose their fear and awe of the forest.[38] This distancing is perhaps what Tebu rejects when he says, contra government perception, that it is the people who live "tame" who kill the forest. By implication, living "wild" is the way to maintain the intimacy of traditional environmental relations.

As this discussion has shown, *həp* is understood at a high level of inclusiveness and abstraction. As an ideal category, the contrast of *həp* is already contained in what *həp* is understood to be. To rephrase, any time the Batek have an image of the ideal forest, they will also have an image of the anti-ideal. When things are going well, the ideal is ascendant. There is no question about what it is and people will find it difficult to put words to their image of the forest.

With increasing endangerment, the forest is looked at in new light; the mirror upon which its image is refracted—the world of villages and towns—grows increasingly expansive. And as that trajectory continues, more and more the forest loses its vitality. The forest seems smaller and more vulnerable; with every new threat to the forest, the world seems increasingly unstable. And thus, the forest's role in the maintenance of the broader world becomes as important as its role in the maintenance of forest society.

In short, ideas that were embryonic under stable conditions become ascendant under conditions of threat. The ideal will be more explicitly enunciated. This play of contrasts between the ideal and the anti-ideal, the valued and the devalued, then has potential as a rallying political force, giving people reasons and justifications for protecting the forest and inspiring ideas about how to do that. Large-scale forest clearance, signaled first by houses-on-stilts popping up in isolated places and then villages expanding beyond the forest margins, is the ultimate advance of the anti-ideal. The body of rituals and cosmological idioms has as its philosophical objective the maintenance of the ideal forest, which is—as Kirk Endicott argued— embedded in a sense of the natural order of things.

Notes

1. Kirk Endicott 1979a, 53.
2. On the thunder complex, see Blust 1981; Dentan 2002b; Needham 1967. Batek reactions to Gubar are discussed again in chapter 7.

3. Critiqued in Carey 1976, 65.
4. Endicott and Bellwood 1991, 160–61.
5. Evans 1937, 11.
6. Endicott and Bellwood 1991, 160 (Table III).
7. Endicott and Bellwood 1991, 161–62.
8. Schebesta 1973, 87.
9. Kirk Endicott 1979a, 68.
10. I thank Niclas Burenhult for reminding me of this.
11. Niclas Burenhult 2002, personal communication.
12. Kirk Endicott 1974, 37.
13. Kirk Endicott 1974, 40.
14. See for example Kent 1989; Nelson 1983; Puri 1997; Serpell 1996.
15. Endicott 1979a, 64.
16. Kirk Endicott 1979a, 43n.19, 56.
17. This point is discussed extensively in Kirk Endicott 1979a.
18. Douglas 1966.
19. Kirk Endicott 1979a, 65; Lye 1994, 276–79; see also Evans 1923, 202–4; Howell 1989a, 181.
20. Kirk Endicott 1979a, 64.
21. Kirk Endicott 1979a, 221.
22. Wagner 1986, 21.
23. Weiner 1991, 38.
24. Descola 1994, 65.
25. Descola 1994, 65.
26. I am indebted to Gérard Diffloth for further explaining the linguistic subtleties.
27. *Kədat* is the Batek name for the poisonous tuber *gadoŋ* (*Dioscorea hispida*). According to this story, the tuber originated from the poisoned body of the cannibal yaʔKədat (who would meet *her* death at the hands of the avenging Paʔ ʔaŋkol).
28. For the full transcription, see Lye 1994, 258–60.
29. Kirk Endicott 1983, 237; Jones 1968, 289–91.
30. See, for example, Kirk Endicott 1979a, 76–77; Lye 1994, 60–64.
31. Dentan 2002b, 165.
32. Dunn 1975, 95; summarizing data collected by E. H. G. Dobby throughout the Peninsula.
33. Whitmore 1997, 156.
34. Karen Endicott 1992, 283.
35. Kirk Endicott 1979a, 53.
36. Whether they are in fact *socially* ordered is a matter for debate. For Java, it has been argued that "it was colonial policy which *created* the peasant village" (Li 1999, 13, citing Jan Breman; italics in original).
37. Dove 1992.
38. Kato 1991, 150.

Plate 8. A camp astride a future swidden field, December 1995. A number of households complained of the heat, and soon after I took this photograph, half the group uprooted their lean-tos and moved into the forested area at the back.

4

In the Beginning

Rattan and gaharu traders have come to mark the passage of calendar-time. Every
three weeks the current rattan trader drives into the forest to pick up his shipments
from a collecting group. On another day, he may rendezvous with a different group
living elsewhere. Whenever he's expected to appear, someone from the group will
walk out to a logging road to meet him and guide him to their latest location. They
might have moved camp since his last visit. The collectors scramble onto the back
of his truck and he drives them to the various wayside points where they've stashed
their bundles of rattan. They load up the rattans, bundle by bundle, on and on,
until the shipments are fully on board. The trader pays them off, maybe chauffeurs
them to the nearest shop where they will stock up on supplies, drives them back to
the forest, makes an appointment for the next collection day, and goes off.

It was on such a day in 1993 that I had my first vivid experience of
deforestation. I was with some men and boys, riding on the back of the trader's
truck as it trundled over a vast oil palm plantation. We were taking the fastest
through-route from the highway to the camp that lay on the other side of the
plantation. The Batek recalled that this area used to be forest; that was not too
long ago. Neat rows of oil palm trees stood there now, a spread of lines and angles
of vegetative sameness. Though shorn of the original floristic diversity and
somewhat corroded, the landscape discontinuities (dips, rises, bends, slopes, gullies,
streambeds) were still discernible. Sometimes there was a break from the visual
monotony; the far-off hills would come into view.

The men were calling out to each other, pointing here, gesturing there. ʔeyTow turned to me. They had just identified the spot where his mother's body lay buried. With the outline of the distant hills as the background visual anchor, they had fixed our position by remembering the layout of the old forest routes. While apparently staring blankly out at the passing scenery, some among them, at least, had been recalling the old associations of routes and topographic relief. There was much consultation. In forested conditions the path is a lot more distinct and the above-ground vegetation provides added landmark features to guide passage. Now, only the most able navigator, ʔeyAlɔr, had the spatial acuity to reconfigure the old landscape in this impoverished space, to accurately recognize the old locations.

As the trader drove us on, ʔeyTow muttered, "What a loss, they have cut down our forest."

There's that word loss (*rugiʔ*) again. Recall its usage from the previous chapter (pp. 52–53) and what this suggests of the commodification and control of the world. Confronted with rows upon rows of an alien tree crop stretching as far as the eyes could see, it is not surprising that ʔeyTow's thoughts should swing so instinctively from his mother's burial ground to external capture of forest wealth. It was not always thus. Before expropriation, there is possession. Before degradation, there is generation. Before death, there is life. Life, the creation of the life process, is the theme of this chapter. Traveling with the Batek has for me been to oscillate between public space and private place, between mourning loss and degradation and celebrating life and vitality. As with "village" and "forest," each informs and defines the other. Both are indelible elements of the landscape now. And beg examination. I focus in this chapter on the origins of the world as told in the myths: the origins of life and the generation and reproduction of its vitality.

The Batek's rich narrative tradition shows little sign of dying out. Stories can teach, often through the recycling of familiar themes, ideas, and problems from one plot to another. In the absence of mechanical distractions, telling stories is good fun. "Set pieces" (as opposed to, say, tall tales, memories, and rumors) include origin myths, etiological and cautionary tales, and folklore. Some are sacred and cannot be told outside the forest, many oft told and well-known. Environmental concerns are prominent; many stories encode useful practical, biological, and ethnoscientific information.[1]

The Batek accept that their stories are true, having occurred in some long-ago time, and are legitimate explanations for the state of the world. Conditions in the world reinforce the apparent meaning of the stories as, reciprocally, the stories render those conditions comprehensible. Forest degradation has given traditional philosophy new meaning. Take the idea that forest cutting destabilizes the land and destroys the world—that's a reverse image of the myth of origin or, in Kirk Endicott's words (explaining what will happen if *lawac* prohibitions are not observed), "deliberate reversals of the creation of the earth."[2] In the myths, after the primordial waters receded from the original disc of land, the creators (see below) called forth the original trees. When those trees are obliterated, the process of creation is negated: the ground subsides, the waters upwell, and the earth is inundated as at the beginning of time.

In their original form, the myths and prohibitions warn against the destruction of the world through human failure to maintain the land/water boundary but forest destruction is not specifically identified as a *cause*. Now, with forest cutting happening at such an alarming scale and reportedly an increase in flooding events, the Batek can see that there is an intricate link between the two. The explanatory framework for that comes from cosmology. To them, cosmology must seem to anticipate current conditions: a case of reality giving renewed vigor to mythology, thus making mythology an ever more muscular system of meaning for understanding and anticipating reality.

There are many versions of these myths. The careful reader of Kirk Endicott's monograph would probably get a different impression of the creation process. So as to preserve the variation between Endicott's and my account, I present my own data in this chapter. Myths and folklore were the topic of my first fieldwork. I was wretchedly persistent to get the narrative flows just right. But I still ended up with loose ends and details that don't quite fit. This is exactly what I should have expected all along:

> You've got to accept the narratives as they are, in all their crazy, fragmented incoherence. It was ever thus. . . . Those with a purist itch will keep looking for some kind of complete, definitive version, but that kind of search is a blatant result of literacy, where the most complete *text* is the version that has the greatest authority. In the woods, there are no texts; just people's memories.[3]

My style of presentation in the next section does not do justice to the thematic variations or creativity in storytelling. For example, in the interests of chronology, I have imposed a sort of temporal order in recounting the myths. But this is certainly not how the myths are told: "there is no 'correct' order for describing the Batek world-view. The Batek never describe it themselves except in bits and pieces, which is also the form in which they learn about it."[4] I would emphasize the performative, dynamic, and "un-textlike" nature of Batek storytelling practices, but this is not the place for it. My concern is to see what environmental principles and lessons we can uncover from the myths. They are part of the average vocabulary of the Batek, told best by men and women in their prime.

The Creation Myths[5]

How was this forest home *made* for the Batek, so that living in there becomes, in Kirk Endicott's words, "part of the natural order of things" (see chapter 3)? This question is implicit in the creation myths. Within my corpus, those detailing the genesis—the initial creation of the world—have shown the most variation. To suggest how emphases may change across tellings, I present some of the versions here.

In Tebu's version (June 3, 1993), the land separated from the waters on its own; he insists that no creator made it happen. That *te?* originally had the texture of beeswax (*sɔ̃t*) and then as time went by cooled and firmed up. Along came the first human couple, a male and a female. Then only came the creators Hala? ?asal and Tohan.[6] Ta?Sudep (Tebu's wife's brother, in his late sixties) then fleshed out the creation image the following day (June 4, 1993):

> In the beginning there were no trees. There was no earth; it was like a *padaŋ* [field]. There were no *cɘmcɔm* [the palm *Calamus castaneus*] to make *hayã?* [lean-to] with. *Batu? karaŋ* [coral] lay under the earth. This was overlaid by a layer of earth. There was water underground, hot water. The land was not hot. Slowly the water bubbled and this hardened the earth. There was the Hala? ?asal and it knew how things should be.

The description is of natural geological process. In the watery beginning, there was no land. But the primordial waters were immiscible. As the waters bubbled and heated, earth and stone matter clumped. Liquids and solids disengaged: liquids became the underground waters, solids grew and hardened into the *te?* as we know it today, the origin of the island that is the world. This is the "heart" (*kɘlaŋes*) of which Tebu spoke in his message (p. 32). Hence were established the first landscape classifications, between hot and cold, between solids and liquids, between land and water, and between above-ground and below-ground.

A third story line was told a month later (July 5, 1993) by ta?Kadɔy. This version begins:

> In the beginning, there was just Tohan and the world. He had all this land but no forest, no people. Tohan thought, "Who will live in this world?" He burnt *kɘmoyɛn* [incense from *Styrax benzoin*] and asked for powers from the Hala? ?asal. *Kɘmoyɛn* was the *?obɛt* [medicine] of the *batɛk hala?*. He made a spell. Then Tohan named the world. Half the world was forest and half was stone. Then was the world complete. It was Tohan who asked the *batɛk hala?* for Gubar [thunder], for *kayu?* [trees], for *batu?* [stone], for *mɘsiak* [humans], for *badey* [wild pig], for *yah* [tiger], for *bɘhɔl* [barking deer], for *kaldus* [silvered leaf monkey], for the foods of the Batek.

In this variant, we glimpse the state of mind of the creator, a creator who poignantly existed all by himself and saw the need for people to live upon the land. So he sought powers to name the world and bring forth its inhabitants and phenomenal features. Unlike Tebu's version, where humans preceded the creators, here the creator precedes all. But this creator, unlike his Judaeo-Christian counterpart, is not omnipotent. For he, too, defers to the Hala? ?asal, the source of his ritual power. Hala? ?asal (or *batɛk hala?* [shaman person]) is a mysterious force, endowed with no gender, no name, no personality, no practical function save to channel powers towards its subordinates. The best translation of the label is "Original Shaman."

There is a fourth variant, and this one is from taʔKadɔy's son ʔeyJudiy (circa July 1993):

> The *labiʔ* [turtle] arose from the sea. After that, there was land. It was old earth, there was *sekepal* [Malay: a clod] of it—enough for just one person to live on it. Lived lived there, the Halaʔ ʔasal, then came another person. There were these two people, one was *batɛk halaʔ*. *Batɛk halaʔ* burnt black *kɔmoyɛn*. The other one was *halaʔgob* [Malay shaman], *halaʔgob* burnt white *kɔmoyɛn*. Then they blew the smoke, blew to the west, blew anywhere. Then *batɛk halaʔ* disappeared. Then rose two people: *tɔmkal* [male] and *yaliw* [female]—Tohan and the Labiʔ.

This version is somewhat similar to Kirk Endicott's: "The formation of the earth began when a huge turtle (*labi '*) rose to the surface of the water at the centre of the sea."[7] ʔeyJudiy's pantheon of creators is slightly more complex than his father's: first there's the Halaʔ ʔasal, joined later by the *halaʔgob*, and this couple is followed by the male Tohan and the female Labiʔ. These are structural opposites—Halaʔ ʔasal/*halaʔgob*, Batek/Malay, black/white, Tohan/Labiʔ, male/female—that map out the fundamental differences in the social world. Like Tebu's, this version plays with the idea that there were two sets of couples and emphasizes that Tohan is posterior, and subordinate, to the Halaʔ ʔasal. But what of the Labiʔ and her functions in this world? That detail is not explained.

Tohan having surfaced, he is left alone to bring forth the world and here ʔeyJudiy's version aligns with his father's in stressing the partnership of Tohan and the Halaʔ ʔasal:

> Tohan worked together with *batɛk halaʔ*. They wanted to bring in the first trees now: *gil* [Malay *toalang*; *Koompassia excelsa*] and *jɔn* [not identified]. These are the two biggest trees, the earliest trees in the world. Got two types of trees.

Creation is explicitly identified with the giving of names:

> He asked asked from the *halaʔ* again, gave names to all these things: this tree, that tree, other trees until he had got all the trees. Then [Tohan] gave names to the rivers, he got rivers. Named the rattan, he got rattan. Named the stone, he got stone. Named the sand, he got sand. He did all this and that's how we got the world we have today. Even now we still have them.

This abundance included food for the first couple whose creation now followed (or became endowed with human needs, depending on the version). To be fully human, they would have to discover how to make babies. This knowledge was conveyed by, in taʔKadɔy's version, the white-handed gibbon (*Hylobates lar*) superhuman being:

> Time passed and the first couple still had no children. Came the *hala ʔ kəbon* [white-handed gibbon superhuman being]. "Why don't you have any children?" it asked them. "Well, we don't know how to have children," they answered. So *kəbon* showed them how. It told the man, "Every month you must copulate with the woman." The couple followed *kəbon*'s instructions and eventually had a son and daughter. This second couple then married one another and they had children. Now there are about five families, two families. When there were about five families, Tohan said, "You cannot marry your brothers and sisters. This is *cəmam* [incest]. Brothers and sisters from the same mother cannot marry each other."

The population thereon began to grow comfortably. At this point, the original creators Hala? ʔasal and Tohan seem to have withdrawn. However, some defects in the creation needed to be rectified. The main problem, as far as the people were concerned, was the provision of food. As taʔSudep said, there were no trees for shelter nor materials for construction. The few original trees had been created. For all practical purposes, however, the forest as we know it today was yet to come and there was nothing for people to eat. It was at this point that half the people sacrificed themselves by becoming trees (see p. 30): they provisioned their friends. The primordial society thus divided itself into two groups, one becoming food-givers and the other becoming food-takers. The close relationship between people and forest was already being established.

As for the people themselves, there had been an error in their making, because they had been given moon-souls (*ɲawa ʔ bolan*). TaʔKadɔy explained: "You see how the moon lives and dies all the time. It never just dies. This was why in those days, people were immortal. They would grow old and die then would become young and live again. How could this work? There were too many people, just like there are many trees in the forest; there was not enough food for everyone to eat." At some point, this problem was corrected (how, when, and by who was not clear to me): the *ɲawa ʔ bolan* were replaced with *ɲawa ʔ pisaŋ* (banana-souls). This was an infinitely better arrangement; as taʔKadɔy continues, "if you look at the banana, when the mother tree dies, we take the young and we plant it somewhere else. The young will grow into a mother tree on its own." That is, *ɲawa ʔ pisaŋ* has the advantage of conferring mortality; as Endicott's informants told him, for both the banana plant and humans, "when the parent dies, the child replaces it."[8] Thus there is an assured replacement rather than limitless growth of the population. A degree of fit between food, humans, and environment was finally achieved.

Something was still missing, though. There was now a forested island and a self-reproducing, sexually conscious, population. But the diet would have been quite monotonous. Along came two culture heroes who brought in the animals of the forest:

Once there were two brothers, they were original beings. One day, they decided to cut a field for planting. They cleared a patch of forest and set it afire. But they had no seeds. Then they went to sleep. Sleep, sleep, sleep, sleep, one of them woke up. He was hungry. "What are we going to eat? We have a field, but no crops." He thought and thought and thought about this problem. Then he had an idea. He cut open his younger brother's abdomen. The blood that flowed he took it and he sprinkled it here and he sprinkled it there, everywhere until it was all gone. As he did so, he put name to the drops of blood and the blood became the crops that were so named. Any-anything—padi, banana, sugar cane, wheat— all originated this way. Only after they had done this, did they go and shoot the *tɔŋ* [masked palm civet; *Paguma larvata*].

Next came the cultivated crops (or, to be more precise, the cultigens):

One day, the two original beings set off to hunt with their blowpipes. They shot a *tɔŋ*. They took their machete and chopped up the carcass. They took portions of the flesh and flung them about. As they did so, they gave the names of animals to these pieces of *tɔŋ* flesh. The pieces became the animals that were so named. The *ʔurɛt* [veins and tendons] became *pacat* [leech]. The *kətəʔ* [skin] became *gajah* [elephant]. The *kəlaŋes* [heart] became *yah* [tiger]. Other animals—*hagap* [rhinoceros], *bəhɔl* [barking deer], *taloʔ* [dusky leaf monkey]—all came into being this way. If the creator beings hadn't done this, there would be no animals today.

Questions like why the pig has a flattened nose, how the sambhar deer got his horns, why the *təluŋ* (probably the bird's nest fern; *Asplenium nidus*) inhabits tree branches, etc., are taken up in other stories. The above (told by Tebu, with *ʔeyTow* supplementing details, on June 3, 1993) are the core myths of origin. They ask two questions: practically, how is the problem of food supply to be resolved and, philosophically, what is the origin of diversity? Both stories depict species differentiating from unitary origins: cultivated crops or cultigens from a superhuman's blood and animals from the body of the *tɔŋ*. The process of creation also roughly follows the same sequence: dismemberment; selection of discrete drops or pieces; flinging them about and the utterance of the names. Once the names were given, the portions took on their special identities. As Endicott pointed out: "the naming of the various plants and animals is an integral part of their creation."[9] The motif of flinging pieces everywhere ensures that the species would be spatially distributed.

As these myths show, the mythical period was a transitional time of genuine fertility, of the origin of habitat, species, and niche. From an undifferentiated cosmos, form and order were laid down. Core landmarks appeared; different plants, cultivated and naturally occurring, came into being; animals were assigned to their names, identities, and niches; the original human population was provided with mortality, consciousness, and a diverse diet and landscape. The termination point appears to be the moment when the Batek became people of the forest. In a moment of mischief I asked taʔKadɔy the question why, if the Batek had invented the cultigens, don't they do any agriculture now. And he proceeded to explain through the *bakar lalaŋ* (burning of the grasslands) myth of origin (introduced on pp. 16–17), which is also in this context about the loss or abandonment of agriculture. (This is obviously the standard answer: reiterated by other people both in Pahang and Kelantan.[10])

To briefly recap, the *bakar lalaŋ* myth begins with a single population of people living together (on or by an island in some versions). Along came a mysterious person ʔadam, the emissary (brother in one version) of Tohan, who set fire to the grasslands. In the heat of the moment, the group that possessed *surat kitab* (sacred books) dropped these and ran into the forest. They got burnt and their hair was singed and thus became the forest people. The other group on the other hand picked up (or stole) the *surat*, then jumped into the water and swam downriver, and they became the Malays. The origin of *baŋsaʔ* (race, ethnic group) is therefore the result of a switch. Endicott has published one version of this myth.[11] I myself heard it at least five times in 1993 and twice more in 1995 and 1996, all from different persons or groups. The 1996 version (yaʔKaw's) came with sound effects. The fire had imposed linguistic order: each *baŋsaʔ* emerged with a distinctive voice or speech rhythm (*kɔliŋ*).

A year earlier, taʔJamal, an old shaman, had his say. Batek came first: the white people are *bel* (younger siblings) to them as the Chinese are *bel* to the white people. It fell on the younger siblings to become rich. In his version, the Kɔrliŋ (Indians) were clearing the field for planting so they set fire to the *lalaŋ*. While other versions suggested that the *surat kitab* changed hands completely in the fire, in this version the *surat kitab* were also transformed. The white people and Chinese took half while the other half were left inside the forest and became the *bab* (food) of the Batek. Obviously, the details of who caused the fire and why did they do so are extremely flexible; the Batek are not obsessed to emphasize who the *real* arsonist was.

That's probably because the meaning of the story is more in what it represents: a pivotal moment in cultural history. It explains how the world that once was became the world that is now and the significance is really in what comes before it: the gradual endowment of spiritual, environmental, and cultural assets. And after—the fragmentation of this abundant world. Until the fire, everyone was Batek and there were no racial differences; the *surat kitab*, however, signified that there *was* an incipient ethnic difference, with one group having literacy and another orality. If we follow the logical structure of the origin myths, and coming after the origin of species, this myth explains not only the beginnings of ethnic differentiation, but how resource environments are now associated with distinct groups of people. This is the final expression of diversity: social and cultural. Finally, the forest acquired a social identity; it became truly the *hɔp* of the Batek.

Of People and Forest

As just recounted, the myths are rich with environmental ideas and references. They emphasize that the Batek's origin is closely associated with the origin of the forest. The image of the forest is strikingly positive. The forest is Eden-like, food-abundant, accessible. It was created for the people and offers many ecological niches that people can exploit: underground, above-ground, land, water, and, less evident here, treetop. Even the sky is reachable by means of blowpipe hunting. These mythical conceptions assert a common origin for all; conflict and strife are absent. A dualistic motif comes up over and over again: male and female, Tohan and the Original Shaman, brother and brother, Batek and Malay, orality and literacy, and so on. The stress is put on complementarity and equivalence; there is no domination of one over another and therefore no subjection and oppression. Only two images might be called gruesome: the disembowelment of the culture hero and the slaughter of the masked palm civet. Even so, these are counterbalanced by what they lead to, namely the enrichment of the forest with new species and foodstuffs. These positive images are certainly congruent with the traditional ecological view of the Peninsula's forests as constituting among the richest and most diverse plant communities in the world.[12]

Where the Batek conceptions depart from some versions of ecologism (for example, in deep ecology) is their anthropocentrism. By this I do not mean that "nature" is subordinate to, or made the instrument of, the expansion of "culture." For example, although we find the appearance of sacred books and knowledge of cultivation in the corpus, we find no myths that explain the origin of material culture or technological development. Rather, I suggest, the environment is coextensive with the world of people. The society of nature coexists with people and is bestowed upon people to make use of. Ultimately, it is the people's needs that are central. When the first couple needed knowledge of procreation, the white-handed gibbon turned up with that lesson. When there were too many people and not enough food, half the people sacrificed themselves for their friends' sustenance. When there was no meat and no cultigens, the superhumans brought them in by supernatural means. People were never abandoned to their own devices; all basic needs were taken care of. The contrast of anthropocentrism is ecocentrism, and this is certainly not valued or even suggested by the myths: "nature" is not treated as a special object whether for devotion or derision. Indeed, nature in the sense of something that is out of culture does not exist.

This point is stressed in the question of origins. The progress of humanity, or human nature, was not straightforwardly linear. There was no evolutionary progress from an original stage of savagery to the heights of civilization, or from animalian to human, leading to a banishment of the wild from the tame. The myths recognize some parallel trajectories. One trajectory was that of people; they were created as they are (albeit with the question of their soul left unresolved initially). Among the animals, some special species (like the white-handed gibbon) may be *sui generis*; they have always existed in their present form and their origins are not questioned or explained. Others, like the species that came from the body of the masked palm civet, were specially created, perhaps to complete the bestiary. Yet other species, featured in folklore outside this corpus of myths, were originally human. The process of creation went in different directions and markedly did not end with a hierarchy where some species or some social groups are left without the qualities that would give them full humanity. Some ideas, like the claim that the Batek preceded the white people and Chinese in the world, could certainly be interpreted as ethnocentrism. However, it is ethnocentrism without discrimination or stigmatization. As shown in the animals, crops, and sacred books stories, what begs explanation is the origin of diversity.

Not surprisingly, trees are a recurring motif, and even celebrated. "Arboricentrism" is common in Southeast Asia (as shown, for example, in taste preference for tree crops [especially fruit trees] and the uses of trees as land, property, kinship, and time markers).[13] It would be very surprising indeed if tree motifs were missing from the Batek corpus. First came the original trees: the first trees created on earth from which, in a way, the whole forest is descended. These might parallel the scientific concept of "keystone species." As ʔeyGk outlines, these trees attract bees, which produce honey, which attracts birds, which pollinate the fruits. If these trees were removed, the birds don't come, neither do the fruits, and teʔ neŋ ʔuhm tahan (the land cannot hold up). Other than their hydrological functions, the trees are important to ecosystem preservation. Cosmologically, they not only keep the land solid, they keep the world going.

There is some variation in the names and identities of the original trees. ʔeyJudiy, as we saw, named the gil (Koompassia excelsa) and the unidentified jən in his telling of the origin myths (p. 81). TaʔJamal named the gil, kəmpɛs (Malay kempas; Koompassia malaccensis), and tagan (possibly the Malay penaga; Mesua spp.). Tebu's list was taduʔ (Oncosperma horrida, a palm), təkɛl (Malay kepong; Shorea sericea), kəpoŋ (Malay meranti; Shorea spp.) and kəmɔyɛn (Malay kemian; Styrax benzoin; a resin-producer). ʔeyGk listed gil, təkɛl, kəpoŋ, and təmiŋ (not identified; but unlikely to be the bamboo of the same name, Schizostachyum jaculans Holttum). Many people mention the təkɛl and some others also add the gil. Kirk Endicott's list from Kelantan was cəmcɔm (Calamus castaneus; a palm), gil, təkɛl, and an unknown species, cinhɛr.[14] Table 4.1 summarizes these lists.

Agreement is not possible at this time. But there are some common characteristics of trees named as "original." Both the palm species have important everyday value: the leaves of Calamus castaneus are the principal source of lean-to thatch while the edible pith of Oncosperma horrida has high food content. They also have woody trunks and therefore in their physical architecture they resemble standing trees. It is also possible that these species are among the most widely distributed woody palms in Batek forest. With the other trees mentioned, height is an important differentiating element: they are among the most discernible members of the tree community, all forming the upper storey of the forest (emergents). In the context of ʔeyGk's remarks, the most important feature about the original trees is that they should be involved in the production of honey and fruits. Their special status is marked in myth. But though these trees have their

Table 4.1. Lists of Batek original trees

Tree species	ʔeyJudih	taʔJamal	Tebu	ʔeyGk	general[1]	Kelantan[2]
gil	x	x		x	x	x
təkɛl			x	x	x	x
kəpoŋ			x	x		
cəmcɔm						x
cinhɛr						x
jən	x					
kəmɔyɛn			x			
kəmpɛs		x				
təmiŋ				x		
taduʔ		x				
tagan		x				

(1) Mentioned in general conversation
(2) Source: Kirk Endicott 1979a, 34

saliency, the forest is not exclusively defined by them. The trees are associated with the production of fruits and it is that whole ecological process that comes together in the definition. This includes the important role of organisms in pollination. Thus, as I think the Batek would agree, what's important is the ecosystem. We should expect this by the Batek's lack of agreement about which trees have original status. Because what matters is the *group*, the fact that all members of the group share a certain set of family resemblances. The original trees are singled out because so much else depends on their persistence and health. They are the indicators of ecosystem integrity and therefore must remain in that ecosystem, reproducing it and being reproduced in turn.

To return to the myths, the original trees were then supplemented by food-bearing trees. This second group was originally human and as such, mark the fact that trees and people share substance, history, and identity. In chapter 3 (pp. 61–63), I mentioned the anxiety over dissolving the boundary between humans and animals. This unease is not—indeed cannot be—extended to plants. The forest originated *because* that boundary was crossed. I would imagine that boundary to be crossed, and the human-plant relationship reiterated, whenever people harvest food to eat.[15] If, originally, people contributed themselves to improve the vitality of the forest, then the continuous harvesting of those plants returns vitality back to the people. As long as this interaction continues, there is a circulation of nutrients linking people and plants in a relationship going back to the beginning of time.

Inevitable, then, that trees should represent, or symbolize, the life process. This is shown in the idea that the soul of humans followed the essence of one particular tree species (the banana) and the parallel concern that having the wrong type of soul (of the moon) would lead to unremitting population growth. This imagery is an interesting commentary on contemporary debates for here we have a long-established, and apparently widespread, aboriginal concern with

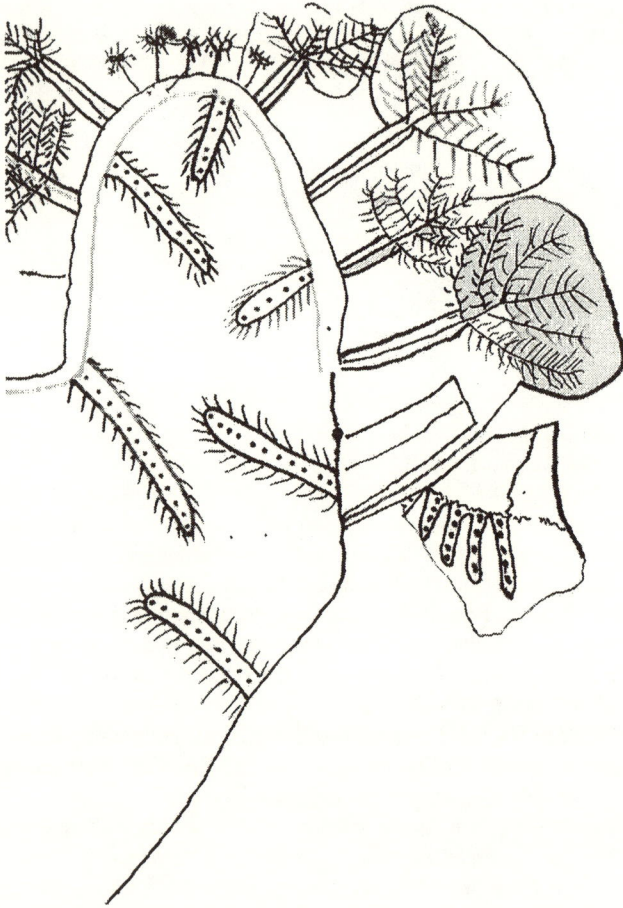

Figure 4.1. A Batek child's image of trees. Anonymous artist, 1996.

overpopulation. According to Skeat, both Semang and Jakun mythologies evince concern that "man at first multiplied so fast as to make the earth too crowded." Outside the Peninsula, an interesting variation on these themes has been documented for the Sakuddei of Siberut island (Sumatra, Indonesia). Schefold reports that one Sakkudei myth has the primordial society becoming worried that "soon the land would be too crowded to feed everybody" so they reached an agreement whereby half the population became forest spirits and the other remained in human form.[16] The Batek image of the forest does include this undercurrent of anxiety about its resource limitations. Tebu's concern that outside claims deplete local access to food is congruent with this. Recall my earlier commentary (pp. 52–53) that contemporary conditions have heightened the sense of the world's vulnerability.

The seeds of that anxiety—or the germ of the idea—is condensed in this origin myth. Where the myth expresses concern with the carrying capacity of the resource base (the phenomenal world of the forest), contemporary environmentalism extends that to the noumenal soul of the world—an intellectual development from the concrete to the abstract, using traditional ideas of the soul.

Tree imagery, in short, is thoroughly woven into culture, symbolism, and, as we saw for Tebu, metaphor. Symbolically, trees, from the root systems to the canopy, visually hold the world together (for a Batek child's depiction of this world, see Figure 4.1). If trees define the forest, the symbolism of trees is ipso facto about the forest: what it is, how its spatial features are cognized and maintained, and what it offers to people. As naʔCaŋkãy said, Batek are people who *"gɔs bah-kiyɔm kayuʔ* [dwell under trees]." Accordingly, a critical function of trees is that they give shelter and, as Tebu stressed, forest clearance deprives us (and the land) of necessary shade (p. 34). Kirk Endicott has elaborated on this: "The main reason the Batek give for preferring to live in the forest rather than in clearings is that the forest is cool. It is not only more comfortable because of this, but also, according to their theories, more healthy. Heat is thought to cause or contribute to many types of illnesses."[17] In line with the general ambivalence, the paradox is that the very things that give shelter and promote good health are also—in the context of treefalls—symbols of danger. This dualistic feature of their view of the forest is shared by other peoples: "while the forest provides the Mbuti with food and other useful resources, it is also the source of disease and other misfortunes. . . . Mbuti regard the forest with ambivalence." The Agta of Northeastern Luzon, similarly, "see the forest not just as a benevolent provider of food, space and raw materials, but also a place of evil, both natural and supernatural (spirits)."[18] Ichikawa suggests, and I would agree, that this ambivalence "gives us a richer image of nature than simply regarding it as a source of goodness only."[19]

Apart from these general messages, the myths also address more concrete intellectual issues. One, as noted, is about the origin of species. The cultigen myth stands out; as Kirk Endicott wondered, it is interesting that these hunter-gatherers should have a story dealing with agricultural origins.[20] The myth gives special symbolic value to blood. Blood is one of the most versatile, primordial, and common of symbols in the world. In this story, its symbolic role is that of claiming genealogy. We can read this as follows: cultigens, that which we associate with domestication and the outside world, are descended from the forest, the body of a Batek superhuman. Indeed, this is how the Batek sometimes talk about food crops, that their origin is Batek. Let me be clear: the Batek are not farmers and traditionally preferred to trade forest products for cultigens rather than to develop their own. But in evolutionary terms, their claim to have invented cultigens is not far wrong (so long as we think about this imaginatively rather than literally). Sure, many Peninsula cultigens are introduced species: prime examples are maize (Malay *jagong*; *Zea mays*) and cassava (Malay *ubi kayu*; *Manihot esculenta*). But many common food crops are improved varieties of forest species—they are descended from their forest relatives, just as the Batek myth claims.[21]

Another dimension is that the myth is right in line with the general theme of the food supply. First the story depicts the superhumans trying out farming. This is a last resort for people who did not have game to hunt, since the animals did not exist yet. So farming is depicted as something hunters can do when they're starving. But they are not farmers and do not practice seed-hoarding and storage; they go through the motions of opening a field but in the end the cultigens must be brought in by supernatural means. Further, the species are not depicted as being under anyone's exclusive control. They are flung about widely. Which suggests they can be "picked up" or harvested by anyone who needs them, and that is exactly how the Batek tend to articulate access to valued resources. The story points to the place of agriculture in the Batek's lives, as a secondary resource strategy that even they can move towards in times of hardship.

The *tɔŋ* story—one of the best known in the canon—testifies to the vividness of animal imagery in Batek thought. What the culture heroes do with the animal's carcass is exactly how game animals are *pə-kritis* (cut up for distribution) in a Batek camp—subdivide, cut, apportion, and distribute. Just as in food-sharing each cut of the meat is accorded a social identity according to where it is sent and who gets to eat it, so too in the myth each cut of the primordial *tɔŋ* acquires a different animal identity altogether. This appears to be one basis (among several) of the Batek's taxonomy of animals.

That each type of animal came from a specific cut of the *tɔŋ* is a clever classificatory device. With the leeches (and other worm-like creatures), the taxonomic basis is one of physical resemblance: these animals look just like pieces of veins and tendons. As Endicott noted, leeches are "vein-like even to the extent of being containers of blood."[22] It is significant that the tiger should come from the heart. As mentioned in chapter 2 (p. 32), the heart is the core of the body and seat of emotions. As the heart is to the body so is the tiger to the forest: it is the most important animal and is what keeps the ecosystem running properly (as shown in one of its names, ?ay Həp [The Forest Animal]). Concerning the elephant, there is a direct parallel: in terms of sheer mass, the skin is the biggest part of the *tɔŋ* and therefore it is the appropriate origin of the elephant. One of its names is ?ay tə-Bəw (The Big Animal); its size is unmissable. The list of other animals mentioned in this version of the story (rhinoceros, barking deer, leaf monkeys, macaques, the gibbons) is illustrative, not exhaustive. They are all large animals—and mammals. There is another classification process going on. The two animals at the top of the food chain, the herbivorous elephant and the carnivorous tiger, both originate from singular (non-replicative) parts of the *tɔŋ*. They are *nonpareil*, in their own class. The severed pieces of veins and tendons and flesh, however, share common origins and characteristics, so from them originate two *groupings* of animals: worm-like and mammal-like. This by no means exhausts the idioms and categories. Still, we now have a nice introduction to the value of species: like trees, they are good to use and good to think with.

Remembering the Landscape

My reason for presenting these myths here is to suggest how the Batek may be conceptualizing degradation. I've taken the section title from Tebu's use of the verb *ʔiŋit* (Malay *ingat*: to remember). As I said, the theme of this chapter is life, or more specifically the origins of life. Mythology presents one version of what the world was meant to be. It was almost too romantic—a world made for people to use and where they were never going to be deprived or abandoned to their own resources. Progressively, that world threatens to be gone forever and thus there is danger of losing landscape heritage as well. Again we turn to Tebu: "*Bilaʔ gən bom, Gubar manah ʔujan. ʔipah ʔiŋit* [When they dynamite, Gubar makes it rain a long time. We remember];" "*ʔiŋit, ʔipah haʔip* [Remember, we miss]" (see appendix A: paragraphs 2, 9, 15, 16, 19). These statements identify environmental degradation as part of the collective memory and also suggest that older memories have a role to play in framing responses to problems. By "older" I mean memory that is channeled through the oral tradition whether as myth, folklore, or just a plain old story. In preserving their oral tradition, the Batek preserve their awareness of the world that was *promised*. Telling stories is one way to slow down social amnesia. I'm suggesting that we consider the mythology as a form of history. Without this alternate vision of the world to draw upon, there is a likelihood that eventually the dominant layer of memory will be all about degradation. Remembering the stories (i.e., having historical consciousness) engenders the feeling of loss, of knowing that something vital is missing. Without this sense of loss, there are no grounds for developing a morality of change: that some changes are just not acceptable.

We saw in chapter 1 how the Batek resent their stories not being heard or respected (pp. 1–2, 17). They link their political marginalization to this. There are, of course, environmental parallels, for what is odd about the Batek is where they live and how they live. As Tebu pointed out, they are often chastised for being "wild;" their way of life is associated with animality (to be elaborated in chapter 5). The forests are socially, economically, and politically undervalued *and* devalued and so are the forest people discredited because of their attachments therein. One objective of Tebu's message is to show us how wrong we are. The mythology takes us some way into this.

The claims that all humanity is descended from the original forest population and that cultigens were invented by Batek superhumans remind us of the links between forest history and "our" history. At the same time that the stories explain the place of people in the forest, so do they, as in the *bakar lalaŋ* myth (and other stories not covered here) try to explain the place of outsiders in the forest: how did these outsiders get here and what are they doing here? Much of that, as I'll explore further in the next chapter, has been about capturing resource wealth. It follows that if the forest has been integral to Malaysian history and development, by default so too the forest peoples. Wrote Schebesta: "They [the Malays] entered the country in search of booty and wealth. In the forest they were more helpless than children.

They therefore attached themselves to the Semang, lived in their encampments, cleared the forest with their help to sow crops, and married their daughters."[23] As this reminds us, forest peoples, and Orang Asli more generally, are not "people out there"—people of no consequence. Within the forest, in the stories that the Batek tell each other, they exercise the authority that outsiders deny them. Telling stories is, therefore, an assertion of strength and a way to claim a history that is often forgotten.

As we saw in chapter 2, the recourse to memory is what enables Tebu to claim that compared to today the past was healthier and more peaceful. Many, like him and ta?Kadɔy, often relate what they know from the cosmology to what they see around them. Tebu's message is an excellent example of how these old ideas are placed in the new context of degradation and of communication to the world outside the forest. Though the circumstances of his communication are dramatic—standing out from the crowd, representing others, entrusting his departing ethnographer with a task—the content, to reiterate, shows continuity with received tradition.

Memory, then, is a source of knowledge and oral history. This is a different kind of history than those inscribed in texts but is no less important for that. The integration of old and new elements that we saw in Tebu's message is becoming a familiar tactic around the world. A number of other studies have demonstrated links between changes in cultural cognition and changes in political economy (for example, transitions to the world system, loss of land and local autonomy, ensuing degradation).[24] New problems need to be addressed within a framework that is locally understandable; often, these analyses are revealed in expressive forms such as myth, belief, rumor, song, and dream omens. As Taussig phrased it, these expressive forms are evidence of "a culture [striving] creatively to organize new experiences into a coherent vision that is enlivened by its implications for acting in the world."[25] In most cases, communities (like the Batek) lack political power and connections. The expressive forms represent local strategies of articulating and documenting concerns and resentments. They reshape the symbolic order of the world, turning conventional notions, such as the inevitable marginalization of forest peoples in framing public policy, on their head.

As shown by this perfunctory analysis, mythology is a rich field for exploration into people-forest and, to the degree that it exists, culture-nature relationships. The stories also articulate a certain understanding of appropriate behavior. Cosmology, ever changing and ever flexible as it is, is a repository of socially shared values and principles that are seen to be "god-given," naturalized,[26] and the basis for the natural order. There is a delicate balance in the natural order and the activities of people can always threaten to destabilize it. Correspondingly, any attempt to invert this established order is conceived as a violation. For example, when you ask the Batek how they feel about the government's planned program of Islamic conversion, they will say this is not right, because the fire in the *bakar lalaŋ* myth decided matters long ago. Malays are the Muslims and Batek are forest people. Similarly, when angry with the Malays, they say that the Malays are descended from them and they are the original Malays (see chapter 5 for more discussion of Batek-Malay relations).

In terms of the conceptual flow of this book, we have now moved from the geographic analysis of the previous chapter to the intellectual origins of Tebu's environmentalism, from *place* to *ideas about place*. We have found that, while the Batek share a history with and are integral to the forest, their approach to the forest is inherently pragmatic and practical. And while the overall image of the forest is benign, there is an underlying anxiety about it, as revealed in (for example) the concern to keep humans and animals conceptually apart, to maintain the separation of land and water, and over food procurement. So the environmental vision is intrinsically ambivalent and it is this ambivalence, I would suggest, that allows for an "ethic of care." In many versions of ecologism and conservation, one protects the forest because of its intrinsic value. In the most extreme versions, there is even an attempt to write humans out of nature because nature is deemed to be more important than people. In the Batek paradigm, one protects the forest because to not do so would be harmful to *people*.

Before concluding this chapter, I should perhaps make some disclaimers. I would not claim that telling stories—putting memories into explicitly linguistic forms—is the only way to preserve and keep memories alive. Neither that personal experience of the forest is the only way to cultivate memory and knowledge of it nor that the Batek are the only social repositories of environmental memories. But perhaps our environmental memories, of a place, of what it means, of the people who have lived there, of the stories about it, of the labor and relationships that have gone into making it, are tenuously held and oft-forgotten. Where the challenge for the Batek is to make their stories heard, the challenge for us is to relate our environmental memories to theirs and reassemble forgotten histories.

Notes

1. Lye 1994.
2. Kirk Endicott 1979a, 79.
3. Campbell 1995, 169, 170; italics in the original.
4. Kirk Endicott 1979a, 30.
5. The myths recounted in this section were collected in 1993, when the language of fieldwork was still Malay. In translating the myths from the Malay summaries and translations that were given to me, I tried to stay as close to the original telling as possible, and this meant preserving details that seemed out of place or chronology. Kirk Endicott's versions of the following myths are found in Endicott 1979a, 33–34, 61–63, 83–87.
6. We have already encountered Tohan in the introduction. The name Tohan is from the Malay word for "god" but this is about as far as the similarity will go. Unlike the thunder-god, Tohan does not have a personality and does not respond in any clear way to the Batek's daily actions. He is most likely not even a god (Kirk Endicott 1979a, 170).
7. Kirk Endicott 1979a, 33.
8. Kirk Endicott 1979a, 85.
9. Endicott 1979a, 63. As Shore points out, "The notion that the origin of the cosmos involves the imposition of linguistic form on an otherwise undifferentiated world is a common feature of creation stories" (1996, 229).
10. Kirk Endicott 1979a, 62.

11. Kirk Endicott 1979a, 86–87.
12. See, for example, Whitmore 1997.
13. Dentan 1991, 423; Dove 1998; Peluso 1996; Tenas 2002. See also Rival 1998.
14. Kirk Endicott 1979a, 34.
15. Kirk Endicott 1979a, 67.
16. Skeat and Blagden 1906b, 183–84; Schefold 2002, 427.
17. Kirk Endicott 1979a, 53.
18. On Mbuti, see Ichikawa 1992, 41. On Agta, see Rai 1985, 40.
19. Ichikawa 1992, 41.
20. Kirk Endicott 1979a, 62.
21. Whitmore 1997, 162.
22. Endicott 1979a, 63.
23. Schebesta 1973, 41.
24. Dove 1996, 33–63; Hoskins 1996; Roseman 1998; Sellato 1993.
25. Taussig 1980, 14.
26. Ohnuki-Tierney has given a clear definition of *naturalization*: "a historical process whereby culturally determined values and norms acquire a state that makes them seem 'natural,' not arbitrary, to the people" (1993, 6).

5

A Sense of Place

Kirk Endicott's 1990 photographs of Kelantan (private collection) show scarred hillsides and valleys where closed canopy forest once stood. Most poignant is one showing Batek men looking from a hilltop over a landscape of degradation. Once before, perhaps, they would never have climbed so high or approached such a view. If they did, what would they have seen? An endless spread of old forest, sheltering the ancient walking route from Kelantan to Pahang (p. 4). Landscape transformation has important visual, and therefore cognitive, impacts. In dense forest conditions, the visual world is normally "close-range, intimate," affording "a sequence of partial glimpses." The words are Gell's, in describing the forests of the Umeda people in Papua New Guinea. As he points out, there is a significant *absence*: "There is nothing to bind all this together, no privileged 'domain-viewing' point, like the view from the keep of a castle."[1]

Other forest visitors have tried to deal with this kind of world. For anthropologist Campbell: "Orientation in the big woods [the Amazon] is like being in a perpetual mist—no horizons, no views, *ever*." For the colonial administrator Frank Swettenham: the Perak forests in late nineteenth century Malaya were "a weird gloomy looking place." Another reaction: "impenetrable gloom." Chapman, the British officer who spent World War II in the central highlands of Malaya, summarized the dissonance thus: "There is always an extraordinary feeling of exhilaration as one leaves the oppressive jungle behind and emerges into the freedom and open skies of the plains." *Oppressive jungle. Freedom of the plains.* Little wonder that some forest dwellers, like the swiddening Temiar of upland

Kelantan or Zafimanry of Malagasy, welcome opened-up areas (in Temiar language, tɛʔ lalah meaning "clear land" is a general expression of approval).[2]

Accustomed—and some would say *adapted*—as they are to closed canopy forest, the Batek also find the "privileged 'domain-viewing' point" a treat; on the occasions when they're up high enough to get a view, they spend hours gazing at the distance. Or, while out walking or foraging, they will pause for the view if the landscape suddenly opens up before them and they get a glimpse of a broader vista. But when the land is shorn of vegetation, this kind of view is thrust upon the people without respite or relief; the intimacy and sense of shelter afforded by closed canopy forest is gone. Thus more than cultural and spiritual values are at stake. It's like being condemned to permanent "visual shock." The sense of place is violently disrupted. In Pahang, the impact has not been as severe as in Kelantan, primarily because the clear-felling came later and the Batek had some twenty years to get used, patch by patch, to an increasingly exposed and fragmented environment. In the Kəciw valleys, for example, they typically foraged in logged-over forest that started to grow back long ago. Resources are less than before, but they exist. This is an altering but not altered world, still replete with cultural meaning.

The forests are rich with biographical and sentimental associations. That's, of course, true for any place, anywhere in the world, with or without forest cover. Yet we find little explicit acknowledgment in many scientific and governmental writings that the forest is also a home where people live, a *place*, and that the people who live there have many attachments to it. Where Batek (and other Semang) are concerned, perhaps one reason why officials do not recognize attachments is the mobile way of life, which seems to evoke a host of "irrational" responses. They are afflicted with the same image as the Penan of Sarawak: "the popular view is that of a people who wander endlessly, perhaps aimlessly, through the trackless depths of the forest in search of food." Almost the exact words were written by Ooi Jin-Bee in an influential textbook of the early 1960s: "The natural produce of any one jungle locality is necessarily limited, so that when this is exhausted the Negrito group will have to migrate to a different locality. For this reason they seldom stay more than a few days in any one place but are more or less continually on the move in search of food."[3] Forty years later, public perceptions have hardly changed, as I myself often discover when talking with other Malaysians. As we'll see below, there are many historical and cultural reasons for this.

The dismissal implicit in this scientific assessment remains apposite: "We have as yet no record of the use of 'high places' or shrines among the pure Negritos, and perhaps naturally so, since the idea of regarding a specific locality as sacred could only grow up with the greatest difficulty among tribes who are so essentially nomadic that they never stay more than four or five nights in a single spot."[4] Which implies that the people only travel in a forward direction, do not find any place worthy of sustained interest and study, are always ready to throw their past away. Whenever place-attachment does come up for official scrutiny, it is dismissed as evidence of *tradition* that should be wiped out by modernity. Gut instinct aside, there are political reasons why mobility arouses such responses; for one thing, it's easier to extract labor and taxes when people are settled in one place. In government-speak: "Orang Asli in this country should change their nomadic attitude so as to be able to enjoy fullest the development efforts of the government."[5] Usually the first stage of government intervention begins by shaping the mode of food procurement: as propounded in the still-current *Statement of policy regarding the administration of the Orang Asli of Peninsular Malaysia* (1961),[6] mobility should not be encouraged and socioeconomic upliftment through adoption of agriculture should be promoted.

If mobility does not militate against having a strong attachment to place and appreciation of history, then what does it mean to *gɔs*, dwell in the forest? I'll turn here to *sentiments*, the bonds and attachments to places. Sentiments are at least as important as practical knowledge to the conception of forest. Expressions of sentiments, contained within open-ended (and sometimes endless) reminiscences, are common staples of Batek social life. We have dealt, if superficially, with the geography of place (chapter 3). The Batek's knowledge in this area is, indeed, an issue of pride with them; they marvel at the range of ways that their society has developed to organize thought and perception. But there is this other body of lore—for lore it remains with them—that explains who they are as people and why they feel so passionate about their contemporary dilemmas.

Longing for the Land

To have a sense of place is to feel abiding affection for it; the word "love" (*sayɛŋ*; Malay *sayang*; appendix A: paragraph 11) comes up often. As with many powerful emotions, this contains the seeds of its own negation: *sayɛŋ* also suggests not wanting to give up or lose that which is loved. For example, someone might ask me for my machete in these terms: *Moh sayɛŋ ba?* (Do you love it?). The intended meaning of the sentence is *Are you willing to give this to me?* The logic behind this word's usage seems similar to those of *rugi?* and *gɔh* (see pp. 52–53); all are words of quite different meanings but arising from a similar history, that of losing place and learning to calculate about it. That is, inasmuch as the Batek love and understand their forest, they also know that the forest is a thing up for grabs, by people whose environmental perceptions are quite different from theirs. Sentiments of place include sentiments of loss and bereavement.

Moving to Bumɔkɔl camp (see map 7) one morning in 1996, we foot-trekked through logged-over forest. We would emerge from forest, cross a loggers' access road, enter the forest again. There was a well-defined pathway that merged with and branched off from the logging road; we were heading for hillsides that the loggers had long since abandoned. We cut through an old encampment. The women with me remembered that their friend na?Makek had given birth there not too long ago. Talk continued; the baby had not survived, was buried elsewhere. Na?Kapey, walking alongside me, frowned and announced indignantly that the loggers later clearcut the baby's place of burial.

Plate 9. Logging roads criss-cross the hills of Kecau. This group of men is setting out to check on the conditions of the 1995 annual fruit crop. On right, ʔeyPaliy leans over to inspect a trailside vine. The tall tree with spreading crown in the background is a *kəmpɛs* (*Koompassia malaccensis*), a common bee-tree. A honey-collecting rope is faintly visible among its boughs.

Most times, the Batek do not voice out their emotions so loudly. Several months later, I went fruit-collecting in Təmiliŋ with a group led by naʔKsɔʔ. It was a long journey, the better part of which spent on the road that fringes the southeastern border of Taman Negara (see Plate 2). We cut through a rubber smallholding, turned upward to the hillsides. A short stretch of road was flattened over, ready for surfacing. Electric poles and cables testified to the imminent urbanization of the area. We sat down to rest, and the reminiscences started. It seemed that the group used to have a camp on that very spot. No traces of it remained. The Batek let this pass without comment; one by one they slipped into silence. I was left to figure out for myself what this means. I tried to picture what it's like being in their shoes and witnessing the gradual erasure of their traces on the land.

Land is, in Read's well-turned phrase, a "country of association and memory."[7] Many times I was struck by the detail in the Batek's memories. Trekking in the forest on routine collection activities, we would come across an old campsite with the wooden frames still standing or, if long gone, the tell-tale patches of young growth once pressed down by lean-tos. My companions might poke around to look for left-behind possessions. They would show me: this is where so-and-so had her lean-to next to the lean-to of that person; that was where that other person did that; they came here from that camp; we had this kind of meat to eat and went fishing there; and so on. If they don't remember, they'll ask others later. The response might come back thus: *They camped there when they did that . . . and that was when I was on that river. . . .* Someone else might chime in, *and that would have happened when I was. . . .* Then someone might strike a different note, telling about something else that happened roughly around that time to another person, who may not be present! This kind of talk always has that context of *place* behind it. Memories are inscribed in the place. As I ended the last chapter, putting memories into explicit linguistic forms, i.e., as stories, is not the only way to preserve those memories. Most important is having the place to return to, for place is powerfully evocative of history. You have to take people there. Going back evokes the most vivid memories.

And affective bonds. The word *ha?ip* (see pp. 30, 32–33) is ever-present, as in these various stories:

Na?Ksɔ? remembers how one night she dreamt that she and a sister were walking by a river, going somewhere. The sister urged her to follow. She refused. The sister went off. When she woke up, she felt inexplicably *ha?ip*. Later the news came—the sister had just passed away in another part of the forest. Kayə? dreamt that something was not right with his favorite brother; he said he felt *ha?ip* and accordingly started to remove himself from the ongoing chat. On another day he said he *ha?ip* the past because there were advantages: hunting was easier and faster, game was more abundant. Kaw, four years old, was throwing a lot of temper tantrums. Her grandmother explained that the child *ha?ip* her mother (whose attention had shifted to her new baby brother). ?eyPa?ic had traveled several hours with his boy Pa?ic (aged about four) to visit our camp group at Tɔm Wal (Malay Wa; map 8). The rest of the family was left behind on the Trɛŋin (Malay Trenggan). That evening, lying in my lean-to, I heard the boy whining to his father. "Why are we here?" The father, voice lowered, gently answered that there were "matters" to look into; I didn't catch the details. The boy *ha?ip* his mother. "Are we going home?" The father replied that they would not, just yet. "I want to go home." The father replied, in the same low gentle tone, that they would stay the night, then see what comes the next day. "I want to go home." Visitors walked in and the discussion was put off.

Ha?ip is central to Batek affect. It involves *longing*: longing for romance, for a faraway place, for a loved one's company. One longs when the objects of such desires are not present or, if present, beyond reach. It may become like an obsession; the Batek say that in *ha?ip* they wouldn't stop thinking about it. It can be evoked by dreams, sensations or premonitions of ill-conditions, memories of past places and deceased friends, upsetting relations, many things in the visual and auditory worlds. It is deeply connected to *nostalgia*. And *desire* and *yearning*. The semantic range of *ha?ip* is quite broad; a close English equivalent might be "to feel moved." Many readers of this book, I suspect, will know exactly how *ha?ip* feels or can identify those sensations, like a favorite kind of music or landscape, that moves them in the direction of *ha?ip*. Pleasures and laments could bring on *ha?ip*. The sounds of a place and time might remind people of other places, other times, and evoke *ha?ip*.

An evocative sound is the call of the Indian cuckoo *pumpakoh* (*Cuculus micropterus*). It is said that when Batek hear the song of the *pumpakoh*, they remember the story of its tragic origins and they feel *ha?ip*. The story is about a Batek man who was cannibalized by Batak. Before he was killed, he warned his wife in any eventuality to follow the song of the *pumpakoh*, which is what he became after his death. It is therefore a love story about how a husband sacrifices himself to lead his wife and child to safety. She, in turn, cannot live without him; she continues to track his call and similarly transforms into the bird.[8] Many other evocative sounds can be heard at *ripat* (dusk), when birds and insects are at full volume. *Ha?ip* wells up strongly then. But, as na?Kapey said, they are "afraid" to dwell on those sounds too long or too deeply. For, as we found in chapter 2 (p. 33), *ha?ip* can lead to fatal illness.

One way to stir out of this state is to do something practical. As na?Tow said, when we *ha?ip* a friend, we can travel to join them. Or when people are nostalgic for the places of their youth, they can journey there. When his wife died ?eyKadɔy shunned the company of his in-laws and kept to the rivers and pathways of his youth. When the sentiment is strongly associated with the place itself, then the common response is avoidance: thus ?eyKadɔy avoided the places (and friends) he associated with his late wife and returned to the places he associated with youthful innocence. When na?Kaw died, her husband ?eyKaw took a different direction altogether. Leaving his children in his in-laws' care, he returned to the escapisms of youth, traveling far and wide with friends till he got to Kelantan. Only later could he return to his parents and the scenes and places associated with his married life. When women talk about their lost children, they say they feel *ha?ip* to see the places where those children had walked and played. Those memories are evoked when the places are seen or remembered. Until the grief has dulled, they can't bear those places.

Much of the everyday visiting and relocation that I observed can be attributed to affect. With movement more places come into experience. Movement itself may be a journey to remembrance and memorialization. In September 1995, I was fishing with ?eyTow and na?Tow. As we walked to the river, ?eyTow pointed to an old trail and said to me, "If you really want to know, that is a trail of the old people. So when people feel *ha?ip* for the old people, they come back here and use the trail so they can remember the old people." If the pathways are gone, it becomes harder to keep in touch with the past. Remembering is an act of sociality, connecting an individual to the vast networks of relations that have led to the present. Pathways have no meaning without knowledge of kinship, moments, events, and experiences told and retold over generations. In returning to those old trails, in making remembering a special event that necessitates practical action, the Batek are affirming their bonds with the old people. As long as they do so, the trails will continue to have meaning and to draw the Batek back to times and places past. They are returning to their history while laying down the paths for the future movements of those coming after them. From this angle, pathways are routes to remembrance and with remembrance lies the chance of discovering more about the land, of having more stories to tell and more stories to remember.

There is, then, a reciprocal relationship between places and narratives. Like with the accounts earlier, going to or passing through a place "calls up" the lore; the people tell the stories to one another, remember the friends featuring in their stories and the things that happened there. They give concrete social meaning to the place, which is as much a part of recognizing the place as the cognitive activity of wayfinding. And the place itself may be well loved for its own aesthetic qualities, pulling people back time and time again, thus setting the scene for more story-making and becoming ever more familiar. " *ʔipah jok, ʔipah ʔiɲit, ʔipah wek* [We move away, we remember, we go back]," said Tebu, criticizing the insult that they are "wild people" who have no landscape attachment. But as landscape deteriorates, the places may be left behind for good. The knowledge becomes cognitive activity alone, bereft of the practical and physical base that gives it continued meaning. The biographical associations, memories, narratives, and conceptual maps are detached from the landscape, becoming history without a place.

Belonging

I turn now to the implications of history *with* a place, as encapsulated in the theme of belonging: the modes and claims of belonging. To belong to the forest is to cherish it, to feel anger at intrusions and irruptions. But it's also to know what and who else belongs. In seeing how this abundant world was created (chapter 4), we glimpsed these other members of the forest world, like plant and animal species and Malays. Some entities belong to the forest, some claim belonging, some just can't belong. I've pointed to traces of an underlying ambivalence about the forest and its many entities and inhabitants. In somewhat related vein, the Batek have a pronounced sense of vulnerability against invasive forces. In the forest, they are at home, the center of their world. But this world is subject to predation. Batek representations of belonging and predation will concern us below. These are part of what I call the iconography of landscape: the images and symbols of forest. In particular, I focus on the two central symbols, *gob* (outsider; stranger) and *yah* (tiger). These images are then contextualized within the frame of the ecocosmology and the Batek's claims that they *jagaʔ həp* (guard the forest). First, a short overview of the history behind their sense of vulnerability. For this, too, is integral to sense of place.

A History of New Arrivals

For a long time now, the forest has been occupied and visited by a stream of new arrivals. Here's a story about that:

> There was a group of Javanese, they were plying a boat upriver. Plying, plying up the river, they saw a beehive in the water. The hive was actually hanging from the branch of a *gil* tree [*Koompassia excelsa*]; what the Javanese saw was its reflection. They thought the bees were *in* the water. "You jump in and get us those bees," the group directed one among them. In jumped the Javanese then: jumped in, bobbed up, over and over again. After a while he took a stick like this one here [gesturing], then he stabbed his friend, so deeply that the shit came out. Then they ate the shit. The one in the water had climbed back into the boat. Then they got hold of the bees and ate these up. Now full, they continued plying upstream in the boat. Their friend, they just left him lying there, thinking he was dead. The friend was starting to smell. Farther upriver, they met up with some Malays who asked them, "Why are you taking this corpse with you?" "Oh, it's not a corpse—it's our friend. He's fast asleep because he's so full from eating bees." Up, up the river they went. The friend was very smelly. The others said, "You must be dead. You stink so badly." They threw him overboard. Thrown overboard, the man was then [washed up] on the riverbank. He met some Malays. "I must be dead." "No, you're not." "But my friends said that I am." "No, you're not dead. It's just that you stink like shit." Took him home and he lived with them. That's why we still have Javanese living here today. The rest, farther up the river they went. Their boat was full of coconuts; these they had brought with them for they intended to settle here. There was a story of fighting at home. They were hungry, but they had nothing to eat. Now they saw another *gil* tree. Thinking to climb the tree to find food to eat, they pulled up and moored the boat. Now they swung down one of the tree branches (*tabiŋ*). They formed a human chain, each man holding on to the one above him. The one at the top, his hands got tired so he had to let go. Let go, everybody fell and died. Only four people were left, and these became *talo?* (dusky leaf monkey; *Presbytis obscura*). So *talok* originally came from Java.

Though clasped within the general problem of explaining the leaf-monkey's origins, this is no doubt a story about immigration. The injured man who's washed up among the Malays is the central figure, the new arrival who's lost all sense of bearing, who, indeed, has no sense of place. He's died and come back to life; he undergoes a rebirth from Javanese to Malay; he makes his identity anew. The story has some nice comic elements, of the foreigners who are so inept in their new land that they can't tell the difference between appearance (reflections in the water) and reality (beehives up the tree), who eat *bees*, who bring enough coconuts to start a plantation but no tools to crack open for the flesh inside, who are clever enough to come so far from home but find themselves stymied by (to Batek) the simple challenge of scaling the *gil*, the tallest legume in Southeast Asia. There's also the predictable dash of ethnocentric suspicion towards these strangers: they're stupid, they're violent, they smell like shit, they don't even know if they're dead or alive.

A story like this invites puzzling questions, like whether the Batek ever got close enough to immigrants to, say, *smell* them or whether the images are derived from other groups' stereotypes. After all, the Batek have usually stayed away from the kind of big, navigable river where the story is set (pp. 56–57). One historical reason was to keep out of slavers' way. The Batek's fear of intruders shows through the Kelsall account. In 1881, Kelsall and the botanist Ridley (who would achieve lasting fame for his efforts to establish rubber in the country) headed an expedition to find a route to the as yet unscaled Gunung Tahan. They went through the southern portion of Batek territory in Pahang, along the ancient trade route from the Təmiliŋ up to the Tahaɲ Rivers (introduced on p. 4). As they went up the latter (in present-day Taman Negara), they observed and made use of the Batek and wildlife tracks on both sides of the river, and also recorded the presence of log bridges over the river and surrounding ravines. However, "[d]uring the whole time we were in the Tahan jungles we did not meet a single Sakai[9] [Batek] although we constantly met with evidence of their presence in the shape of rough leantos. . . . Other evidences were small platforms in the forks of trees, 15 or 20 feet from the ground, dead fires and on one or two occasions newly cut branches of trees." From these, they concluded that "[t]he wild men are evidently very shy, as they never let us catch a glimpse of them although they evidently watched us all along, and on our moving from one camp were quick to clear off all old tins and other waste which had been thrown away."[10] The image is telling. As scientists and explorers advanced, forest peoples moved back—a nice metaphor for the appropriation of forestland. The Batek were fearful of the colonials' bodies but plainly not of their material culture.

External interest in the forest, and the state-making process of which it is a part, has been thoroughly documented. The slave-raiding period which lasted until the start of the twentieth century is the most salient in cultural memory: "What is perhaps most interesting is that the *memory* of slave raiding has been preserved by most Orang Asli groups, and this exerts an important influence upon their actions today. . . . [S]tories of slave raids and similar abuse at the hands of outsiders form an important component of the oral tradition of many Orang Asli groups."[11] Highlighted for the Batek are images of slavers ambushing forest peoples to capture the women and children for the slave trade (dubbed *gob ʔamɔʔ* [Malays went amok]). Slaves could be sold to raise cash or made into cheap domestic labor.[12] In the Peninsula, true slaves (as opposed to debt-slaves) were Orang Asli. In 1878, when the Russian explorer-ethnologist Miklucho-Maclay encountered Batek in Lebir, Kelantan, he was told of two subgroups. The terminology is significant: one group the Malays labeled "Sakai Liar" (Wild Sakai) and the other "Sakai Jina'" (Tame Sakai). The former lived deep in the forest, kept their independence from the Malays, and reportedly even "retaliated" for the abuses they had suffered from them (he does not, however, detail what form this "retaliation" took). The latter lived close to Malays, had socioeconomic relations with them, and even intermarried with them.[13]

As the terms show, the Malays had a rich mythology about the forest and forest peoples too, and this can be placed within the larger context of immigration and the genre of *conquest narratives*: "The role of tribal peoples in the mythical charters of several Malay states has frequently been noted. The idea of a dynastic marriage between an immigrant ruler and an autochtonous tribal woman . . . is not uncommon in the Malay world."[14] There is legendary evidence that in earlier centuries settlers acknowledged their debts to Orang Asli: Orang Asli were invested with magical powers, recognized as the true aboriginals of the land, incorporated into some state administrations, and up until the nineteenth century some Orang Asli groups played a role in the selection of Malay leaders. Their mastery of the forest was their economic niche. They were the acknowledged experts and the primary collectors of forest products.

New arrivals to the Peninsula lacked this knowledge of the land. For them, the forest was on the one hand a spatial and spiritual obstacle to human settlement. Thus we find that, even though "The Malays have no single, simple division between culture and nature," they had a range of practices to propitiate spirits (of water [river and sea], earth, and forest) for human intrusions into their abodes.[15] As Kato comments, "human contact with the forest, whether in the form of felling trees or collecting forest products, always required a ritual mediation." Maxwell describes the classic Malay perception: "The Malays always consider themselves as intruders when they enter the forest, and never forget their awe of and reverence for it. They seldom go into the forest alone. . . . While it is true that the forest lies almost at their doors, they never forget not merely that no man knows its extent, but that it actually is without bound or limit."[16] In the Malay classification system, common among farmers, the forest is "out there" and people are "in here." The idea of "intruding" on the forest is critical; it implies deep respect for the forest's powers and acknowledgment that it is a whole other world in which one might get irretrievably lost. In other contexts, the forest (as is common around the world) could be considered a source of spiritual renewal and regeneration, a spatial "Other."

At any rate, *respect* is an alternate face of the widespread Austronesian tendency to represent nature as a "wilderness" that needed to be conquered, defeated, subjugated, or "tamed" before dynastic belonging could be assured.[17] The advance of colonial interests in agriculture and mining (symbolized by explorations like those mentioned here), which stimulated the rush to clear more and more forestland, was the endpoint of this process. In the modern period, the Malay perception of forest became increasingly predatorial (at least at the upper circuits of capital). Once the wilderness was "tamed"—the landscape reshaped in a commercial and administrative model (p. 73)—respect for its other-worldly qualities would have faded. The nineteenth century saw a growing marginalization of the forest and subordination of its peoples. Orang Asli were irrefutably relegated to subordinate status and viewed with contempt and in some contexts, as accounts like Miklucho-Maclay's show, with fear. It is a continuous process of social and cognitive distancing.

As Malays psychologically "retreated" from the forest, their store of folk images may have become more pejorative. The Malay civil servant and traveler

Abdullah bin Abdul Kadir offered this revealing impression in 1849:

> The first thing I noticed was that in their general bearing they were human beings like ourselves, but that in their habits they were hardly as animals. For animals at least know how to keep themselves clean, which the Jakun [of Malacca] certainly did not. . . . Their eyes had a wild look in them as though they were ready to bolt. As they chattered to one another they sounded to me like birds twittering.[18]

In every sentence, the Orang Asli are described *as* animals. Such images lingered. In the mid-1920s, Schebesta recorded: "The people living around the dwarfs [Semang] relate among themselves very extraordinary things about these tribes, but always in whispers, lest a dwarf hear. The story goes that they are not men at all, for they suddenly bob out of the ground in the most unexpected places; that they have glowing eyes, are ignorant of the use of fire, and eat everything raw."[19] To be fair these images are by no means confined to the Malays; the Batek, for example, also have similar stories to tell about *other* peoples who are half-human cannibals. Descriptors like Sakai Jina' and Sakai Liar underscore the general idiom through which Malays saw the Batek. *Jinak* (tame) and *liar* (wild) were and remain common Malay descriptions of patterns of Orang Asli behavior and are obviously the source of Tebu's bitterness, as expressed in his statements *"People who live tame, they kill the world. JHEOA officers ridicule us for living wild."* Jinak represents the kind of behavior that is most compatible with Malay ideals of social relations and *liar*, with its animalian associations, is the exact conceptual opposite of *jinak*. The two terms Sakai Jina' and Sakai Liar encode what local Malays saw in Orang Asli and how they classified different behavioral patterns. But there may be more to it than that; as various authors have suggested, investing subject peoples with qualities of the exotic wild, or negative stereotyping more broadly, makes it easier to enslave them.[20]

One thing seems clear. Semang were not shy of confirming Malay prejudices about them:

> the Malay reasons that the Negritos, being to all intents and purposes animals, must, therefore, be in league not only with the wild beasts, but with all the supernatural and inhuman powers of the forest that he dreads so much. On occasion, therefore, and individually, he is somewhat chary of offending the Negritos too deeply. The Negritos, on their side, too, sometimes are not backward in taking advantage of Malay beliefs in their regard.[21]

"Taking advantage of Malay beliefs" is a good example of a broader Orang Asli strategy to accommodate themselves to these intruders, to protect lineaments of their "autonomous way of life,"[22] or to get them "off their backs."[23] Miklucho-Maclay again: "The Malays . . . are very much afraid of these Orang Liar [Lebir Batek] and do not venture either alone, or in small parties into those parts of the forest which they are known to frequent."[24] Thus, by cultivating their "fierce" image and capitalizing on Malay fears, the Batek would have kept their territorial space.

The contest for land was undoubtedly a major force in all of this. As Malay polities stabilized under British rule and expanded their control of the interior, Orang Asli in remote fastnesses made perhaps their last-ditch attempts to repel intruders. Clifford's 1895 documentation of the Tem-be Sakai (Temiar) of Nenggiri, Kelantan, shows much affinity with the Miklucho-Maclay account: "These tribes are said to number several thousand souls, and as they bear a bad reputation among the local Malays, the interior of the Nenggiri district is almost entirely given over to them, very few Kelantan natives ever penetrating far into this Sakai country." So belligerent were the people ("these Sakai have frequently committed depredations on Malays entering the district") that raiding parties would be sent by order of the Sultan "to keep the jungle people in check, and to punish them for their misdeeds."[25] Those Orang Asli who could not fight back learnt to use what Scott has called the "weapons of the weak": among them, "foot dragging, dissimulation, desertion, false compliance, pilfering, feigned ignorance, slander, arson, sabotage."[26] "Dissimulation" (giving a public face of congeniality), "feigned ignorance" (acting stupid) and "false compliance" (pretending to obey) best capture some of the Batek's style of communicating with outsiders today. It is telling that Scott's theorizing of the weapons of the weak was based on fieldwork among peasant Malays in Kedah: i.e., as much as Orang Asli may have felt oppressed by Malays, so did poor Malay peasants feel oppressed by their rich neighbors. One effect of dissimulation is to confirm and reinforce the outside perceiver's biases. The word "shy" or synonyms thereof is an endemic adjective in colonial-era accounts of the Semang and Orang Asli are still afflicted with the popular image of being simple folks with harmless ways.

Reading between the lines, and taking additional insight from Batek lore, it seems that slave-raiding did not create a *paralysis* of fear. When conditions were peaceful, some Batek groups could and did venture out to visit, live close to, and work for upriver Malays, camping close enough to them to develop a range of socioeconomic relations—and observe visitors coming up the river. Some of the Sakai Jina' may have been slaves or former slaves, or *were considered as such by Malay patrons.*[27] But when slavers (often "henchmen" from downriver[28]) tore through the forest, those Batek who could, fled. Concurrently, at any point in time some groups were willing to be more exposed to villagers and others kept contact to the absolute minimum. From Batek discussions, it seems that though they generally stayed away from hostile confrontations, groups in the past *were* capable of aggression.

By the 1930s, Batek groups in the Taman Negara area (by now generalized as "Pangan") may have won tolerance by local Malays. A small trace of this is in the three-volume report of the Wild Life Commission headed by T. R. Hubback.[29] The Commission's mandate (1931–1932) was to investigate the conservation situation at the time and propose solutions to the problem of balancing wildlife preservation with crops and people protection. Hubback was also asked to investigate the possibility of excising land to set up the first national park (now Taman Negara). The territorial implications of the proposals were quite obvious to everyone, as shown by testimonies in the report. Aborigines are mainly represented through the sympathetic observations of others. The Batek ("Pangan") appear as informants advising Hubback on a stretch of the Pahang-Kelantan border and are mentioned by a Tembeling Malay headman thus: "The Pangan merely wander about and do no harm to anyone or anything."[30]

A terse remark and still redolent of ingrained biases ("merely wander about"). Nevertheless, in view of the conservation politics of that time, we can read this as a shrewd attempt to protect or, at least, not jeopardize the Batek's interests. As Kirk Endicott has suggested of the slave-raiding period, upriver, rural Malays may have found it necessary at times to protect Semang—if not disinterestedly— from lowland incursions.[31] Keep in mind that the Batek are hunters *par excellence*. A conservationist might not have wanted them in a future national park and wildlife sanctuary. One of Hubback's inspirations was the Yellowstone model in the United States and there, as is amply documented, indigenes were expelled from their lands in order that "wilderness" areas could come into being. Today, the access rights of local peoples to protected areas remain a contentious issue throughout the world.[32] Thus, by casting the Batek as little more than animals, the Tembeling headman was reiterating that they belong to the forest and need not be messed around with. Indirectly, the "wild" image was serving the Batek's long-term territorial interests.

If not paralysis, then a fertile and lasting guardedness. Among Semai, for example, "as late as 1962 many Pahang Semai would leave riverside settlements for the security of the forest when they heard the sound of a boat coming upstream, only to emerge when it was clear that they were safe."[33] Anthropologist Rodney Needham had a taste of the Kelsall and Ridley type of experience in 1956, when he went to the same area expressly to look for the Batek.[34] By then the Emergency battles had crested. Insurgents were largely routed out but stray members did find their way to Orang Asli communities, some in fact using them to hide out from authorities. And, as I said, the Batek didn't want to get involved with these newer types of intruders (p. 49). Needham was led by the Chewong guides Beng and Patong, who had been brought to work in Taman Negara by the Game Warden C. S. Ogilvie. They would have been superb trackers. Yet all they found of the Batek were remains of camps, fireplaces, dug-up tubers, and the like. Needham did not reach the conclusion but when I read his account I suspected that the Batek were just ahead of the party; they probably knew they were being sought and made themselves scarce.

According to received history, fear stopped the Batek from exposing themselves to intruders. One memory is that the young men would be sent on shopping expeditions by themselves. Upon return to camp, they would split up, each person taking a different route, meandering round and round so as to throw pursuers off their tracks. Around the time of Needham's visit, a young ta?Jamal found his curiosity of the white people getting the better of him and—so he told me late one night in 1995—he urged his friends to scope them out. He hid behind the bushes, fearful yet drawn closer and closer, watching the rangers build some of the earliest wildlife hides in the park. Subsequently he too fell under Ogilvie's spell and was trained to a laboring job in Kuala Lipis, there to see for the first time its main street, its Chinese shophouses, its restaurants, its bus terminus.

By the early 1970s when Kirk Endicott visited Taman Negara there was an established Batek settlement at Was Tahaɲ, right in the grounds of park headquarters.

From Stranger to Friend

This history of having to make space for new arrivals, and fleeing outside attacks on their bodies, has done much to shape Batek perceptions of vulnerability and belonging. Wariness is bred early. As Karen Endicott observed: "when children must be stopped from a dangerous activity, parents must resort to frightening their children by telling them a tiger or a stranger will get them if they continue their behavior."[35] Right from the start of life, children learn to associate these two entities with malevolent forces that they should recoil from. As Guemple notes for the Inuit, such tactics "assert the presence of threats from outside the circle of community members . . . [and] underscore the boundaries of the solidary community . . . by pointing out that outside there are only 'animals' . . . and 'enemies.'"[36] If Malays represent everything strange to the Batek, tigers represent everything they fear. Tigers and strangers are structurally "top of the heap," symbols of authority, and belonging yet not-belonging to the in-group.

As we've seen, Malays have had many roles in Batek society: as immigrants and settlers who coveted forestlands, employers, customers for forest products, patrons, "owners," neighbors, raiders, traders, government representatives, and so on. Malays are important actors in this place and their behavior and attitudes toward the Batek have varied on an individual and historical basis—by turns friendly, helpful, well-meaning, kindly, tolerant, generous, superior, abusive, possessive, patronizing, bossy, threatening. If they, especially those from the interior areas, did not have positive qualities to recommend them, the Batek could not have approached them. Most times, relations are (at least visibly) amicable and friendly. But it is based on political and economic inequality and the Batek see the Malays not only as the "Other" but as a superordinate people who look down on their ways of life.

Still, despite all this, there is evidence of close interaction. For example, ʔeyMaŋkap remembered how as a boy he got very curious about the Malays so he left his family and stayed in a Malay village for a while. ʔeyTow remembered how there was a Malay woman who adored his children and every day came to take the children to play in her village, where she fussed over and spoilt them thoroughly. Ladiy identified a Malay family on the Təmiliŋ whose children's ages are parallel with her children's and how the two sets of families were childhood playmates. NaʔAlɔr said there was an old Malay woman living outside Kuala Lipis who used to observe the Batek in the old days and harbored many recollections of them. Reciprocally, Batek can reel off the kinship relations of many long-established

Malay families. All of this testifies to a letting-down of suspicion and distrust, mutual participation in local histories, and a rich exchange of intercultural information.

The memory of abuse and slave raids does remain powerful. In times of tension today, the negative side of the relationship will flare up. There are many Malays who continue to fling insulting words like *sakai* (slave or dependent; Batek *sakey*) at Orang Asli. Every new insult and pressure from Malays confirms Batek biases about *gob* and the rightness of not letting the guard down. I observed one attempt at pressure in 1995, when a Malay man on his motorbike, recognized as an avid promoter of Islamic conversion for Orang Asli, stopped some Batek youths and harangued them how they should not "*lari* [Malay: run]" all over the forest like animals, how they had to settle down in one place and become developed. The use of *lari* as a synonym for Batek *jok* (to move [residence]) is obviously dismissive and hints of a deeply entrenched ideological bias. There are more polite Malay terms, like *berpindah-randah*, which is closer in meaning to *jok*. This was not an unusual instance.

Since about the 1970s, government attention has increasingly focused on assimilation through Islamic conversion (in officialese, "spiritual development"). In view of this official position, the Batek generally cast government attention as a way to "*taŋkap* [Malay *tangkap*: to catch]" them; i.e., entrap them into adopting Malay modes of being. One could say that this guardedness, sense of being at the butt-end of Malay schemes, provides a fertile space for developing ideas about cultural integrity and continuity, identity, and ethnicity. Malays are good to think with; as Dentan famously titled his essay, *If there were no Malays, who would the Semai be*.[37] Ideas about the Other are devices to highlight ideas about the ideal Self.

In chapter 3, I suggested that the world of the village is the mirror through which ideal-type images of the forest are refracted. Batek environmental representations show evidence of a deeper need to "make room" for Malays in their scheme of things, to figure out whether Malays belong to the forest and the limits to that belonging. My way of tracing this is to look at the transition from *gob* (stranger; outsider) to *kabɛn* (kin; friend). One who is not *kabɛn* is by definition *gob*. The symbol of *gob* are the Malays, the archtypal outsiders, while *kabɛn* are strongly associated with the forest, the world of insiders.

At its most particular, *gob* refers to the Malays. Malays as an ethnic group represent agrarian life, aggressiveness, sedentariness, and Islamic modes of thought and behavior. When the Batek say that one of themselves is behaving "*macam gob*" (like Malays), it is this complex of ideas that they refer to. The more general meaning of *gob* is *any* stranger, anyone who is not-us. Fieldworkers who take up long-term residence in the community thus enter as *gob* and, if successful, leave as *kabɛn*. Because of the pejorative connotations, the Batek are not anxious to admit that the word carries this broader range of meanings. For example, back in 1993 I tried to confirm whether a non-Malay like myself is also a *gob*; I was assured that I was not. Yet a moment before I had heard Kidey (the future ʔeyKapey) call out to a friend to look for *kasut gob* (*gob's* shoes; i.e., mine).

An important component of identity is having *kabɛn* in the forest. Most commonly, *kabɛn* are relatives from birth and marriage but there are connotations of strong friendship, bonding, and companionship: *kabɛn* are preferred traveling and working companions. The Batek clearly abhor solitary life; to be an orphan (*bə-kəpok*) is a condition to be pitied. The network of *kabɛn* is the community that one belongs to. But this community, as I remarked over the nature of groups (p. 11), must have porous social boundaries. Otherwise, it would be an infinitely closed society, which it has not been historically or genetically.[38] Hence there are elements of extended kinship in the *kabɛn* concept that would permit the Batek to confer (if fictively) a Batek identity on *gob.*

I'm arguing that a *gob* is always a potential friend. The transition from one status to another, from "not-belonging" to "belonging," is hedged by pre-established conditions. Without going into the social dynamics and ramifications of this process, it is worth considering the role of the environment in this process. To become *kabɛn* is not only to belong to the forest but to be accepted by, and ipso facto become *part* of, it. When a stranger moves into camp, the initial reserve of the Batek hosts is more than compensated by the thunder-god Gubar; either that day or soon afterwards the arrival of the stranger is marked by thunderstorms. I was told that Gubar registered a most clamorous complaint the night after my first visit to the Batek. The natural forces, in other words, objectively express what the Batek, in their social demeanor, cannot directly show: their initial sense of disquiet. (It is expressed in body language and a shyness with face-to-face contact, but the Batek do not turn strangers away. Verbal insults are practically taboo.) The stranger's arrival, then, is fraught with risk for the Batek. To become *kabɛn*, the stranger should adopt the norms and practices that obtain in the forest. Most relevant here are those pertaining to the ingestion of substances: eating forest foods.

As we saw in chapter 2 (p. 29) food is an important metaphor, not only of social relations but ethnicity. Dietary issues become pressing when strangers stay with the Batek. The latter eagerly probe their guests about their food preferences, usually framing the questions in a comparative context and offering analyses that would do an ethnologist proud. Most pertinent, they want to know if the guests have any dietary restrictions like the Malays' Islamic prohibitions. A persistent refrain in 1993 was, "Batek don't have any food taboos. Malays have many." (The former is, however, not strictly true.) Of special interest is meat from the hunt; the Batek's faces positively glow when they talk about game foods. This is precisely that part of the forest diet considered *haram* by Islam. The Malays' food prohibitions effectively limit the extent to which they could adopt Batek practices. Since they cannot partake of wild game, they symbolically have to reject the forest from which that food comes. As such, they will remain the archtypal *gob.* So long as *any* stranger resists participating in food-related norms and practices, they are not *kabɛn* and cannot belong.

From Predator to Prey

To belong to the forest is also to know of its dangers. Kirk Endicott is right that the Batek "do not fear the forest,"[39] if we take this to mean the forest in the abstract, in the sense in which it is perceived by agricultural peoples. The Batek's fear is a different sort: fear of what is known to be possible. Belonging is also, as I said, about knowing what else belongs to the forest and being sensitive to its potential for both good and bad. Trees are a good example of the environment's capacity to evoke both positive and negative responses. As we saw (chapter 4), the origin myths celebrate and put high value on trees. Yet, trees also bring a great deal of unease. Nothing is so terrifying in the forest as the sudden collapse of a giant tree (p. 56). Symbolically, to dream of tall trees is to receive a warning of floods upwelling from below ground. To be killed in a treefall is to the Batek a horrifying prospect, equivalent to death by tiger-attack, interpersonal violence, and a fatal fall (as marked linguistically). This is part of the poignancy of belonging to this place: to live, prefer to live, under the forest canopy and have the knowledge of what a fragile environment it is. This is precisely the kind of knowledge that fosters the ethic of care that I've introduced earlier (p. 93).

The Batek-Malay relationship, as we saw, is based on unequal access to resources and reflects social-economic stratification in Malaysian society more broadly. What of stratification *in* the forest? We can find this in the Batek's relationship with animals. From a subsistence perspective, this is the relationship of hunter and game, or predator and prey. But there is an important cognitive and symbolic side to this relationship. Earlier chapters have noted a distinction between big and small animals, with elephants and tigers representing the former (pp. 60, 90). The tiger (*yaḥ*) is pre-eminent; it immediately leaps up as a worthy counterpart symbol of *gob*. It's worth an extended examination here. For one thing, it violates accepted categories. Further, as the chief predator, its fate is symptomatic of the animal kingdom as a whole.

The tiger is a magnificent beast and eloquent of much cultural meaning. The Batek view it, and other predators, with fear, circumspection, curiosity, and the occasional mockery. To use Guenther's terms (writing of animals in Bushman art), the tiger has "aesthetic salience." Around the world, animals of this class "spring to the eye, presumably any eye, irrespective of cultural conditioning. They command a high degree of aesthetic and cognitive impact."[40] There's supporting evidence of this in Batek identification of animals. With large species like tigers and elephants, there is universal consensus about the names. From one identification exercise to another, all respondents unerringly put the correct names onto the pictures in my books. These animals' physical morphology are clearly etched in their minds, unlike those of smaller (and actually more commonly seen) species.

In Southeast Asia generally, the tiger is commonly associated with "kings, ancestors, shamans, and magic"[41]—in a word, potency. "Top of the food chain," "flagship species," "king of the jungle" are associated images that roll tritely off the tongue. To the Batek, it's also, less respectfully, "Earth Digger" (Cikok Teʔ) and "Smelly Claws" (Haʔac Kalkok), to honor its habit of marking scent with paws and, apparently, because its hands are smelly when it's dragging food around. Tiger fortunes rise and decline with those of the forest and it is therefore an appropriate symbol of forest. It is often claimed that tiger numbers in the Peninsula have declined from an estimated four thousand in 1957 (the year of Malaysian Independence) to the present five to six hundred and that this latter figure is half the number remaining of *Panthera tigris corbetti* in Southeast Asia. With forest loss and increasing urbanization, most people in Malaysia have never seen tigers outside a zoo. Though the tiger remains the central icon of local wildlife, popular knowledge of tigers, including the metaphysical and spiritual aspects of knowledge, has gone into retreat. On the ascendance is the tiger as conservation object.[42] We'll return to this below.

As reference markers, tigers and Malays have equivalent properties. If Malays symbolize strangers, tigers symbolize predation. The tiger is never far from Batek thoughts when they talk about the problems of living in the forest. If *gob* virtually means "Malay," *bĕc* (one of the tiger's eleven names) is synonymous with "predator." Both are threatening to the people in the forest. They stand on the boundaries of Batek society and are yet integral to social order, with Malays representing everything "not-like-us" and the tiger representing both the worldly and other-worldly sides of the forest. The commission of certain classes of proscribed acts, like *dos*, is said to bring a punitive visit from the tiger; the Batek are subject to the limiting conditions of the forest. But here is the difference: Malays come from outside the forest, tigers from inside it. Thus, Malays symbolize the *dəŋ*, and tigers the *həp*. As we have seen (p. 90), one of the tiger's avoidance names is ?ay Həp (The Forest Animal); it defines the forest. Indeed, as shown in the myth of animal origins, it is the "heart" of the forest. Further, since the Batek are people of the forest, they cannot do without the tiger, for without it the potency of the forest would be lost. Thus, Guemple's mapping of "animal" onto "out-group," cited earlier (p. 108), does not quite apply to the Batek. If the Batek have no concept of a wild disorderly "nature" that has to be "tamed" through productive labor, so do they lack a concept of the tiger as a pest, marginal and distanced from society.

As in their relationship with *gob*, and in line with the general Southeast Asian perception, the Batek's relationship with the tiger is fraught with ambivalence and ambiguity. *Gob* may be alien, but are providers of goods and jobs. They also have the symbols of wealth: things and property that the Batek in some sense covet. *Gob* are not disliked for what they are (essential ethnic qualities) but for their rejection of the forest, as shown in their obvious distaste for the Batek's food (behavior) and their tendency to want to control the Batek (politics and economics). As for the tiger, Endicott explains:

> On the one hand, [the Batek] greatly admire [the tigers] for their awesome strength and speed. These qualities make the tiger a very appropriate symbol for the notion of superhuman power on earth, which is personified in man by the shaman. But the Batek greatly fear tigers as well. The identification of the shaman with the tiger seems to be a way of converting the danger of tigers to power for man.[43]

A number of Batek tales posit the problems of were-tigers: sometimes a human is revealed to be a tiger in disguise while at other times a tiger longs for human relationships and assumes human form to realize it. It is one thing, as with the shamans, for the Batek to appropriate the tiger's power for benign purposes; it is quite another when the tiger turns that power against people. The general problem is that the boundary between human and tiger societies is slender indeed and I suspect that there is some deeper horror at the prospect of this boundary crossed for evil purposes.

In its evocation of threat and danger, the tiger may be equivalent to Malays. But, as Wessing has pointed out of Southeast Asian symbolic orders more generally, the tiger's "animal nature and human nature co-exist within him";[44] it has human-like qualities. As an integral member of the forest community, the tiger is the mirror image of the Batek themselves. The tiger shares many characteristics with them. It occupies their preferred habitat; it travels along waterways, loves to play in the water, and belongs to the lowlands and foothills. Belonging to animal society, the tiger properly belongs in the game category and yet it preys and is not preyed upon. Eating tiger meat is not unknown but hunting tigers is simply not done. The tiger is never game or commodity to the Batek. Why the tiger (and most of the larger fauna like elephants, bears, rhinoceros, seladang [wild cattle], wild pigs, and so on) is not hunted remains to be determined. One reason is that they are rare on the ground and their numbers have depleted greatly since the start of the twentieth century; today, they are more often spotted close to farming areas than in mature forest. When their numbers were more numerous, the situation might have been quite different.[45] At any rate, the general tactics for surviving the tiger—as with all predatorial forces, including treefalls—is to stay out of harm's way; i.e., to practice avoidance and deterrence.

Avoidances take us beyond predators, to a larger complex of environmental relationships. This is particularly clear for linguistic and symbolic avoidances. Avoidances mark out that shifting terrain where animals, plants, and humans interact and roles occasionally get flipped over: food procurement and processing. We have seen how the myths highlight the naming process; i.e., that to utter a name renders an object conceptually concrete and publicly knowable (pp. 80–81, 83). But names have the power both to create and desecrate, to bring both the wanted and unwanted to one's side. One common tactic among hunters is to use avoidance names. As shown in the Batek label *kənmoh pənəwak* (lit. name of disguise), avoidances are a form of secret language whose meaning is known to the human utterers but not to the animals. A common theme is that the name of the animal being hunted cannot be uttered from the start to the end of a hunt. Before the kill, uttering the name will alert the prey to its oncoming fate. After the kill, uttering the name might provoke revenge, or show disrespect to the animal. Thus for example "[t]he Ulu Telom Semai . . . have an extensive set of nicknames which they use from the time the hunters or trappers sight an animal until the animal's meat is digested. Speaking the animal's [name] might lead to severe gastrointestinal disorders or even death."[46] In effect, the names relate to the power over life and death that hunters wield over their prey and the need to constrain this power with a system of morality.

While Batek name conventions are not so elaborate or scrupulously observed, their behavior shows traces of the same kind of unease. For example, when a carcass is brought to camp (if you like, crossed the threshold from "nature" to "culture"), those at the receiving end cannot mock the animal's name or do things like point to it or comment unfavorably on it. The main comment or question is "*yaliw ba? təmkal* (female or male)?" Not much more is said beyond getting someone to butcher the carcass and distribute the various shares. If the animal is interesting enough, people say "*yɛm tɔt* (I want to see)" but they tend to observe in silence or talk about matters other than the carcass in front of them. Even children observe, though not so obediently, the conventions: "so this is what it looks like," a child seeing a species for the first time may say. Commonly they stare wide-eyed. (However, they do allow fieldworkers to photograph or weigh the animals before processing.)

The tiger, the archetypal symbol of predation, naturally has the most number of avoidance names to its credit. As ?eySətsɛt explained it, they'll use the avoidance names when in the forest and most vulnerable to attack. They are afraid that if predators hear their real names, they will think they're being called. In relation to the game animals, the Batek are the predators. In relation to the predators, they are the prey. So they cannot blithely call out the names, for who knows what assault that might invite? The use of avoidances shows respect for the predators' awesome powers over man as well as good caution. However, avoidances also exist for non-predators (game animals) like the rhinoceros hornbill (*Buceros rhinoceros*), dusky leaf monkey (*Presbytis obscura*), and giant frog (*Rana macrodon*) and even some species of wild yams. As a way of relating to predators and prey, the use of avoidances acknowledges that the species are sentient: can hear, can understand, can respond.

In some contexts species are like humans: endowed with internal organs, the usual complement of head, hands, feet, ears, facial features, the capacity to move and procreate, and have relationships with members of their own kind. And, even supposedly inanimate species like the wild yams (*takop*) may have features in common with humans. For example, the Batek do not say that yam roots "grow" (*suroŋ*) in this or that direction; rather the roots *cip* (walk). And when a root doesn't want to *cip* anymore, that's when you get a stub. Then when it changes its mind in mid-course, it *cam* (looks for) another place to *cip*. These plants have fairly mobile tendencies, just like humans and animals. More than simply have a mind to move along a path, however, the root may have the capacity to apprehend the harvester's intentions as well: when a dig fails, the yam is deemed to have *talak*, the general purpose verb to denote any kind of flight or escape—another example of the tendency to cast forest relations in terms of predation and intentional avoidance of predation.

But species are also not like humans. For one thing, they are a food source. Many prohibitions (see pp. 61–62) assert the importance of keeping animals in their place. Cooking prohibitions like those forbidding the mixing of incompatible categories of food over the same or different fires extend this caution to the world of plants and bought foods.[47] Humans and nonhumans inhabit separate but complementary niches. Animal captures and plant extraction cross these categories, turning social species into dietary sources; i.e., they change their identities. Avoidances return the cosmos to its proper order, by symbolically denying the real identity of the species so captured.

But, as I said, the tiger is the central icon of Malaysian wildlife. As such, tiger perceptions outside the forest are worth noting—and, in view of the larger threats to tiger survival—perhaps more important to trace than the Batek's. Here are some heated words from Game Warden Kitchener some forty years ago:

> A curious anomaly regarding the tiger is that, on the one hand it is worthy of inclusion in the crest of the Federation of Malaya, on postage stamps and a multitude of trade marks and yet, on the other hand, amongst the many forms of Malayan wild life, it is given a status to if not lower than that of the wild pig, rat and squirrel and treated accordingly: to be destroyed on sight by every possible means.[48]

Slightly further back, we get some thinly disguised anthropomorphism from Government Administrative Officer A. Locke (ridding Terengganu settlements of predatory tigers from 1949–1951): "I have rarely shot a tiger without a pang of regret that another courageous, strong and graceful creature has died. To me they will always be by far the most fascinating and magnificent specimens of our Malayan wild life." The habit of thinking of tigers as worthy opponents has been identified as a common one among colonial hunters: in India, Greenough reports, there was a "cult" of the hunt, which included "an unwritten code insisting that sportsmen must court danger in maintenance of honor and must recognize the essential nobility of the hunted animal."[49]

Perceptions vary, of course, depending on person, time, and situation. The politically correct response these days is something close to pity: *the tiger is doomed to extinction if we do not protect it; it is not to blame for all the crimes (of attacking humans) that have been charged at it.* Tiger perceptions are class-differentiated with those living close to the forest, who feel the menace of tigers more intimately, taking a more practical view. As I composed these passages (2002), tigers had hit the newspaper headlines once again following a series of attacks on smallholder farmers by "man-eating" (usually displaced and malnourished) tigers, three of them fatal. Probably the most controversial response to tigers that year was offered by Nik Aziz Nik Mat, the chief minister of Kelantan where some of the attacks occurred: "They should all be shot. Malaysia already has far too many tigers. They are better off dead." The chief minister's statement provoked a furore in the press but may have proven the adage that bad publicity is better than no publicity. Among the spinoffs, the World Wide for Nature Malaysia (WWFM) has found that it's become easier to raise funds for their tiger conservation projects.[50] Even without intervention from politicians, the tiger is eminently mediagenic.

Without photographs and films, I'd be at a loss; despite living in the forest for months on end, I have never seen a tiger walking free in its natural habitat. Ironically, villagers are more likely to encounter tigers than forest dwellers. By clearing forestland in tiger territories for crop production, they cut into habitats and enrich food supplies with new browse.[51] Accordingly, they are virtually calling the tigers to them. In the forest, tigers and people are a mutual avoidance society. This is not to say that the Batek never encounter tigers or have not seen people picked off by tigers; only that under normal circumstances they give the tigers a wide berth and practice deterrence. They recognize, for example, that one deterrence is campfire smoke. If there are no camps in a place, it is considered *jɔm* (cold), becoming *saŋyɛt* (empty of people) or *haɲiw* (quiet). Then the tigers will come. Tigers are known to *kitʔɔ̃t*, that is, to circle round an area when they smell smoke. By keeping track of everybody's residential histories, people like taʔKadɔy (who has an avid interest in intercamp communications) are monitoring the landscape, assessing which areas are empty of people or likely to be tiger-infested.

Of course this is not an infallible system; tiger behavior is patterned but tigers also shift territories in response to landscape changes that may occur far from the Batek's purview. Nighttime, when many animals are awake, is most dangerous. Once in 1993 Tebu's group woke up to find a line of tiger tracks right through their campsite. They had slept through the tiger's visit. There is the amusing "true life" story of the colonial surveyor in Negri Sembilan who, "finding the fire unusually warm, he nestled more closely against the fragile wall. . . . [In the morning] the poor surveyor realized that he had spent the night cuddling up against a tiger for protection from the cold."[52] There is a clear need to be vigilant at night. I was scolded once for not sounding the alarm when I heard the rustle of a four-legged animal moving through the shrubbery; I was the only one still awake (in my defense, I retort that I was performing the anthropological duty of studying the sounds *and* my campmates' capacity to sleep through such disturbances).

By valuing avoidances (cognitive and practical) and practicing deterrence, the Batek are furthering the aims of conservation. The conservation problem arising from a man/tiger fight for space does not apply in this case. The Batek and the tiger inhabit complementary but, as far as possible, not overlapping niches. Kitchener, cited above, rightly states that human beings are not normal prey for tigers: "If this were so the aborigines of our forests would long ago have been exterminated."[53] As the game warden, Kitchener had a duty and responsibility to set the record straight. However, the problem is not whether people understand tiger behavior but rather how particular interpretations of ecology serve social and political ends or, conversely, how conditions bias the choice of interpretation. For example, as the top politician of the state, the chief minister of Kelantan obviously had to come out on the side of people, even courting controversy to demonstrate his sympathy for the plight of settlers. As social scientists routinely argue,[54] there is a distinction between event (the actual magnitude of the threat) and representation (the discourse of threat). How people think about tigers is influenced by how and where they live and what their roles in society are. Kitchener does not, for example, consider that Orang Asli avoid "extermination" because they draw on a good sense of geography and wildlife behavioral patterns. That knowledge is part of the sense of place, which is absent from newly arrived settlers.

Changes, Moralities, and Claims

Earlier chapters have examined landscape concepts (chapter 3) and then ideas about and origins of landscape (chapter 4). Following on that theme, this chapter has discussed senses of place in the context of sentiments, lore, history, symbols, knowledge, and relations with other members of this variegated community. Those "others" were Malays or external claimants more broadly and animals as symbolized by the chief predator, the tiger. The intent is to reveal what it's actually like to live in this landscape, the relationships among people, between people and place, between insiders and outsiders, and between people and biological species. A collateral theme has been about belonging to this landscape. That includes dealing with relational risks: those coming from people who are not-like-us and with animals that are like-and-not-like-us. Organizing these relations is a certain code of conduct as signaled by a range of taboos and behavioral constraints touched only briefly in this book. Within this world predation is a given. But now there are new givens. The upper layer of this chapter's story is about what is being lost—landscapes of association, memory, and belonging—and to signal how relationships might be changing. For example: the loss of places to return to, the process of cognitive and social distancing characteristic of the Malay trajectory, the gradual objectification of tigers seen in the larger Malaysian society, the reshaping of landscape history.

Chapter 4 described how human and natural history belong to the *same* "natural order." In line with a common theme in the anthropological literature, the objectification and distancing of nature is taken to be somewhat suspect. It is useful to think of people and forest (that is, all the biotic life associated with it) as constituting a *moral community*. The community coheres because there are ongoing relations with the superhuman beings. To abuse these relations, performing proscribed behavior, could bring on ecological danger.[55] Without stopping for a more detailed examination, we can find an underlying theme, which is that the prohibitions are about *respecting* the natural order of the world. There is a morality, in short, that minimizes the risk of human actions disrupting the relational complex of the forest. This fabric of relationships could not exist *sans* the superhumans, who are protectors, providers, and "moral watchdogs." Taking the metaphor a step further, we find social qualities invested onto different members of this community. This was shown in the discussion of avoidances. This is not to deny that natural entities have objective qualities independent of their sociality; rather, the sociality gives them an added identity.

Perceptual change is part of this story. The Batek state that the forest would collapse if they, its primary dwellers, did not observe the moral codes. Which suggests that they see themselves at the center of the forest, their activities integral to its continued life. As Kirk Endicott writes, they acknowledge that they have "a well defined part to play" in the forest.[56] But when the Batek go on to say that they *jagaʔ həp* (guard the forest; introduced on p. 50), this suggests something quite different: that the forest is a "thing" vulnerable to misuse and/or predation, and therefore subject to calculation and *distanced* from themselves. This tension between opposing impulses can be traced to landscape changes.

The exact provenance (to time, place, and individual) of the term *jagaʔ həp* is unknown. We find no evidence of it in Kirk Endicott's writings. Reflecting its recent origins, it does not seem to have become a full-blown ideology, replete with theories of causation and prescriptions for behavior. It still consists of a vague series of assertations, articulated with varying degrees of conviction by different individuals. I accept that it is on the way "in" and potentially has moral force; as such it is worth discussing here, albeit superficially. Its meanings and implications do show continuities with the older cosmological idioms. Kirk Endicott pointed out of the Kelantan Batek in the 1970s: "Although individual Batek may leave the forest for varying lengths of time, it is generally believed that if all the Batek moved out of the forest, the superhuman beings would destroy the world."[57] This belief is still relevant and indeed was among the earliest themes that I encountered in my own fieldwork. But now there is the *jagaʔ həp* layer superimposed upon it. The belief can be restated as follows: "if all the Batek leave the forest, there is no one left to guard it and the superhumans would destroy the world." And other forces come into play, notably the pressure towards sedentism and Islamic conversion—neither of which is compatible with forest-dwelling. Thus what was once an abstract issue has become concretely relevant.

Jagaʔ həp was one of the first phrases I heard among the Batek in 1993. Something happened between, say, 1980 and 1993 to bring this hitherto implicit idea into the realm of public discourse. Those years saw a growing build-up of logging in Kelantan, which was to result in clear-cutting and plantation establishment. We know that in the late 1980s and early 1990s the DWNP was worried that a recent spate of logging in Kelantan would drive too many Batek into the national park and disrupt the ecosystem there. Well they need not have worried. Whether deliberately or not, a "system" of turn-taking developed in short order. At any one point in time there are only on average (let's say) 150–250 people (perhaps rising to three hundred at peak season) inside the national park. By the time I took up fieldwork in 1993, all of this had stabilized and the uppermost issues discussed with me were forest cutting, dam-building, and land collapses. The context for asserting the claim to *jagaʔ həp*—to a kind of spiritual stewardship—is absolutely contemporary. As conditions exacerbate the Batek's sense of threat and vulnerability, their notions of the forest as lifeworld are increasingly complemented with a notion of forest as abstract political, economic space.

To render the notion of stewardship more concrete, they cast it in oppositional terms. Most particularly, they single out the Malays. *Gob* kills the world, said Tebu. "They only care to get rich," says ʔeyGk. They point out how they, the Batek, never clearcut the forest, only *bɔt* (take) what they need, and strive to live within ecological constraints. They argue that there is a difference between their dwelling practices and the Malays' increasing urbanization and promotion of monocropping, resource commodification, land transformation, and industrialization. This is a commentary on the problem of scale, from the point of view of a minority people who are constantly on the edges of that development, feeling its impacts, and thinking of long-term consequences. The conceptual division is here established as one between "us," people who know the forest best and how to take care of it, looking out at "them," people who are ignorant and care only to take as much as they want.

One problem for the Batek is that all peoples properly should observe the behavioral codes of the forest: the beings are omniscient everywhere, even outside the forest. But outsiders, whether in their own territories or visiting the forest, are not emotionally tied into their cosmological rules. The Batek assert that when people come "up" to the forest, they should observe the norms that obtain therein. Otherwise, they endanger themselves, the forest people, and the forest. Variations on this theme are often reiterated by different individuals, with or without my probing. It has become a general grievance. The poignancy of this assertion is that only in the forest would they have any confidence that they are at the center of the world.

One last word. I've said that the forests are also defined in relation to the world outside (chapter 3). It has a political identity. Not only is that world—*our* world—defined by its different biophysical properties (as shown in the discussion of forest-village distinctions), the people of that environment are also known for keeping their distance from the forest. As often noted, farmers are typically concerned to keep the boundary between village and forest as clearly defined as possible. Rambo has suggested that farmers cannot live in the forest without radically transforming, and ecologically simplifying, it.[58] Considered thus, the forest is de facto land where the indigenous trees are not yet cut down, and it can be guarded by the Batek because no one else has claimed it *yet*.

Notes

1. Gell 1999, 241.

2. Campbell 1995, 25; Aiken 1973, 142, 145–46 on Swettenham; Chapman 1997, 311; Geoffrey Benjamin 2000, personal communication on Temiar; Bloch 1995 on Zafimanry.

3. See Feld and Basso 1996 on local senses of place; Carey 1976 on irrational responses to mobility; Brosius 1986, 173 on Penan; and Ooi 1963, 177 on Negritos.

4. Skeat and Blagden 1906b, 167.

5. *Mingguan Famili* [Family Weekly] 19th November 1989, cited in Zawawi 1996b, 581.

6. Jabatan Hal-Ehwal Orang Asli (JHEOA) 1961.

7. Read 2000, 22.

8. For the complete transcription, see Lye 1994, 270–72.

9. During the colonial era and right up till the 1950s, Sakai was the popular term labeling all aborigines. The Batek were later placed in the sub-category of "Pangan," which refers to the Semang of the eastern half of the Peninsula.

10. Kelsall 1894, 45. For a comparable experience, with the unrelated Batek Tanum, see Wells 1925, 129–30.

11. Kirk Endicott 1983, 237; see also Benjamin 2002; Jones 1968, 289–91. Recently Robert Dentan has suggested that Semai fear of thunder and more generally their sense of cosmic malevolence is related in part to "living on the edge of a state based on slavery" (2002a, 224).

12. On the various tasks performed by slaves, see Kirk Endicott 1983, 216–17.

13. Miklucho-Maclay 1878. This is the earliest concrete ethnography of the Batek. Miklucho-Maclay's account is curious in one respect. Though his vocabularies and description of Batek movements prove that he did encounter the people, his classification of "wild" and "tame" is obviously derived from the Malays. In other words, he may have observed the Batek in the forest but he seems to have got much of the interview data from the Malays rather than the Batek. For the broader commercial context of Miklucho-Maclay's expedition, see Cant 1973, 26–7; Skinner 1878.

14. Benjamin 2002, 46.

15. Endicott 1970, 111 and *passim*.

16. Kato 1991, 150; Maxwell 1982, 7–8.

17. As reviewed in Schefold 2002.

18. This is a well-known extract from the much reprinted and translated *The Hikayat Abdullah* (1849), here reproduced from Nicholas 2000, 70.

19. Schebesta 1973, 13.

20. See, for example, Benjamin 2002, 48–49; Dentan 1997. See also Woodburn 1997.

21. Evans 1937, 32.

22. Nicholas 2000, 77–78.

23. Benjamin 2002, 49.

24. Miklucho-Maclay 1878, 213.

25. Clifford 1992, 104–5.

26. Scott 1985, xvi.

27. Laidlaw 1953, 155; Skeat 1953, 112; Skeat and Blagden 1906a, 577–78. See Kirk Endicott's (1983, 226–27) review of Malay claims to "own" slaves or subjects.

28. Noone 1936, 55.

29. Wild Life Commission of Malaya 1932.

30. Wild Life Commission of Malaya 1932 vol. 1, 319.

31. Kirk Endicott 1983, 227.

32. See Cronon 1996 on the American construction of wilderness ideals.

33. Dentan 1991, 436.

34. Needham 1976.

35. Karen Endicott 1981, 3.

36. Guemple 1988, 137. Similar tactics are employed by the Javanese (Geertz 1989) and the Chewong (Howell 1989b, 49), among other Southeast Asian peoples.

37. Dentan 1975.

38. Kirk Endicott 1997.

39. Kirk Endicott 1979a, 8, 53.

40. Guenther 1988, 195.

41. Wessing 1986, 1. See, for example, Skeat 1901, 26–27; Schefold 2002; Wazir-Jahan 1981, 86–87. Compare Greenough 2001, 151–52.

42. For a parallel process, see Greenough 2001 on India.

43. Kirk Endicott 1979a, 141.

44. Wessing 1986, 116.

45. For attempts to address the hunting question, see Kirk Endicott 1979b; Rambo 1978; see also Locke 1993, 49–51.

46. Dentan 1967, 101.

47. For more discussion, see Kirk Endicott 1979a, 73–76; see also Dentan 1979, 36; Howell 1989a, 182, 205–6; Needham 1967, 272, 277, 279; Schebesta 1973, 189; Skeat and Blagden 1906b, 223.

48. Kitchener 1961, 202.

49. Locke 1993, 3; Greenough 2001, 161–62.

50. Reported in the *Far Eastern Economic Review* November 7, 2002.

51. Kathirithamby-Wells 1997; Wharton 1968.

52. Keyser 1993, 271, 272.

53. Kitchener 1961, 204.

54. For example, Fairhead and Leach 1996, Knight 2000.

55. See also Kirk Endicott 1979a, Chapter 3.

56. Kirk Endicott 1979a, 82.

57. Kirk Endicott 1979a, 53.

58. Rambo 1982; see also Hutterer 1985.

Plate 10. Malay trader laying out his wares. Photographed by an anonymous Batek youth, 1996.

6

Gathering in the Forest

One of the metaphors used by Tebu was the Malay expression *cari makan* ("looking for food"); in his case to point to the appropriation of forest wealth (p. 29; see also appendix A: paragraphs 3, 18). I suggested that he contrasted this to the indigenous *cam bab*, which does not have the metaphorical connotations of the Malay expression. Underlying his critique of the search for wealth was a basic grievance: that food supplies in the forest *have* gone into decline. Hunting is said to be less productive than before; game is less easy to find. As discussed in chapter 5 (pp. 115–17), there *have* been drops in wildlife populations; the tigers' shifting patterns of predation—changing territories; eating humans as a food substitute—symbolize the general condition of forest resources today.

This chapter continues with a major theme of this book: loss. We've examined different dimensions of loss: of the world (chapter 2), forestland (chapter 3), environmental memories (chapter 4), and place (chapter 5). Much persists, however. Lest I give the impression that the Batek are sounding the final note to a long history of foraging, I want to offer in this chapter some concrete examples of how the forest continues to sustain them. How, that is, they continue to *cam bab* while others *cari makan* from the forest, activities that may be less productive than before but have the possibility of continuing in some modified form. I've said that both life and death, presence and absence, beg examination (p. 78). Chapter 4 described the generation of this abundant world. Elements of that abundance can be gleaned from the Batek's use of plant resources. I cannot offer a methodical outline of Batek botany here. Rather, I provide a general description of the knowledge and harvest of two sets of resources, the yams (the traditional starch staple) and fruits (the Batek's favorite foods). The symbolic context of animal predation was a concern of the last chapter. Now it's time to turn to the vegetal domain, but in a more practical vein. I examine how the Batek use and affect the composition of species. My objective is to give a flavor of the forest's resource and biological richness and, concretely, how the Batek satisfy their gastronomical interests. I hope in this discussion to make a case for the importance of plant resources to Batek health and survival and to document Batek practices in maintaining those resources.

Wild Yams

The generic term for wild yams is *takop*, which covers more than one genus (see appendix C). It is generally assumed that wild yams were the main sources of carbohydrates before the introduction of cultivated plants.[1] For the Batek, *bay takop* (to dig for yams) is a central cultural symbol. This seems to be reflected in children's games, which often feature the theme of digging. Such games either around the vicinity of the camp or next to a yam-digging parent start at around the age of two; these develop motor skills, eye-hand coordination, stamina, and muscular strength within a general context of play and fun. Even babies younger than this will stab the ground with the point of a knife or machete while their mothers are hard at work digging up the tubers. By the age of six, young diggers can become most absorbed in what they're doing; by seven they may have joined in plenty of games featuring "real" as opposed to play-acting tuber-digging; and by ten all children have seen their parents dig countless times and know how to identify the most important yam parts. A twelve-year-old is capable of independent procurement of tubers.

No doubt, Orang Asli foragers have a sophisticated body of knowledge about yam species. It is likely that they underutilize this knowledge. Kuchikura's foraging studies of Ulu Trengganu Semaq Beri (traditionally hunter-gatherers) include the most detailed of Orang Asli yam ecology so far and can be used for comparative purposes here. He reports that the Semaq Beri mainly collected eleven species of *Dioscorea*, compared to the thirty-two species names they had given him[2] while my equivalent data would be six species out of the thirty-three or so listed in appendix C (excluding the varieties or subspecies and avoidance names).

Yam Growth and Morphology
Some background information may be useful here. Holttum describes the growth cycle of the Dioscorea:

> Yams are monocotyledons: their stems do not increase in thickness as they grow older. The stems are slender, and they climb by means of twining round the stems of woody plants. Each yam stem grows from a tuberous base, and it takes

Plate 11. The transmission of yam-digging skills. *Top:* Kaw, then aged three, standing by while her mother and grandmother take turns to dig, April 1996. *Bottom:* Three years later, she has matured into a competent digger and graduated to an adult-sized digging stick.

Table 6.1. Plant parts of yam species: Summary of terms

Plant part	Total terms	Species specific	Growth stage specific
flower	1		
leaf	1		
root	9	6	
stem	6	1	1
tuber	5	3	1
winged capsule	1		
Column Total	23	10	2

food from the tuber during its early stages of growth. As the stem becomes older, its leaves manufacture an excess of food beyond that needed for the future growth of the stem itself, and this food is conducted downwards and stored in a new tuber (or tubers) which grows at the base of the stem. After about ten months the growth of the new tubers is complete. The climbing stems have done their job and they die.[3]

The Batek's classification of yam growth and morphology is quite detailed. The careful labeling suggests long-term cognition and intellectualization about the yams. From a utilitarian perspective, these terms seem exaggeratedly detailed. For example, they were not used in teaching me how to dig. I would not have uncovered them had I not picked out all the discarded pieces from naʔAlɔr's various digs and asked her if they had labels. As we saw earlier, learning is through practice and experience. Children's recognition of yam growth and morphology develops mainly through close involvement in their parents' lives, imitating what they do, rather than through explicit verbal instruction. Thus a child need not master the terms in order to develop digging expertise. Nor are detailed labels necessary for most everyday purposes, one possible exception being that of *pəntɔn* (communicating availability and location information to future harvesters). Out of that context, an experienced yam-harvester can use sight, touch, taste, and for some notable species even smell to interpret the growth process of the plants.

I have thus far collected twenty-three terms for yam morphology. As analyzed in Table 6.1, these terms focus mostly on the root, stem, and tuber; most terms are general but some pertain only to species that share distinctive morphologies, and two are specific to growth stages. It seems that all terms except the general ones for leaf (*haliʔ*) and vine (*tənat*) are peculiar to yams. The labels key into those above-ground parts that signal the root's presence (vines, leaves, flowers, and for seed-bearing ones the winged capsule from which the seeds expode) and therefore the appropriate place of digging. Among these, the most important traces are the entwining (*bit-bɛ̃t* [to be entwined]) of the climbing vines. Another apparent function of the labels is to indicate edibility, i.e., which parts of the root lead into or between the tuberous portions and which ones signify the end of the root's travels through the ground.

Kuchikura's comment applies equally well to the Batek: "The Semaq Beri are well acquainted with the growth cycle of the yams and adjust their collecting activities to it. They never cut the stems, nor dig the spot close to the base of the stem to avoid damaging its stalk."[4] Concerning growth cycles, thirteen labels have been recorded thus far. It has not been possible to obtain definitions of all the labels. Obviously a key criteria of all stages is the stage of edibility, which is correlated with the appearance, including color, of the above-ground leaves and stems (the parts that need to be identified before a dig commences). The labels follow roughly the order suggested by Holttum's summary earlier, but with one difference: the Batek continue to label growth stages up beyond the life of the climbing stem. Some stages in this group are *jïmkil*, when the leaves are brown and drooping, the vine is all dried up and twig-like, and the tuber is starting to *soʔ* (rot); *saŋ* when the tuber is aged already but still contains enough moisture to be eaten, and for *wɔh* species *lakɔk* when tubers have been on the stem so long that they can't be eaten.

Glaŋgadiŋ stage when the vine is browning (i.e., has conducted most of the food into the tubers) is reckoned to be the best stage to dig at. Contrastively *hapɔn* is a long stage before that when the tubers are still *bɔlhɔt* (watery) and this is the stage *not* to dig at. It is not clear to me whether such indicators are in fact used to guide digging practices. For example, it is possible that a harvester might chance a dig despite negative appearances "just in case." In between stages, when the yam is in transition from partially to fully edible and vice versa, or when some tubers on the stem are ready for eating and others not, it may be a matter of instinct or preference whether or not to dig up that plant. Such decisions only become absolute or categorical in food-marginal situations, i.e., during starvation periods when there's a need both to fill an empty stomach and to conserve body energy, and these did not occur during my fieldwork.

The language used to talk about yam growth is not necessarily esoteric or specialized. The adaptive explanation for this is that the terms are easily learnt, unlike specialist terms in any scholarly endeavor that require long years of special study. This would make sense, in view of the yams' overwhelming importance in diet, nutrition, and cultural meaning. Another feature of the language is to point to similarities between animals and yams. Such terms include:

- *dut* (the inedible stub of *pam* tuber's edible roots = avoidance name of the pig = the type of nose associated with the pig);
- *gɛl* (fleshy tubers = lower back of human body = side or part);
- *ciŋ* (to grow horizontally [of tuber roots] = to straighten the legs = to cut a straight gash on a tree);
- *pi-wɔʔ* (to cut off the fleshy tubers of *wɔh*, *pam* and *kɜnsey*, which grow in bunches off the main stem = to cause to rise up [of human body]);

- *cəribuŋ* (to grow slantwise down, derived from *buŋ* [to dig a long, deep tunnel] = an avoidance name of the giant frog *Rana macrodon*, ʔay Cəribuŋ Lənti? [The Slanting Tongue Animal]).

Availability and Digging Patterns

Karen Endicott's 1975–1976 study showed that yams are available year-round, although the rates of return were highest in October and May.[5] Other than the articulation of yams with rice (see below), digging for yams also shows a seasonal pattern: the Batek do not dig for yams when fruits are available. This statement was said most vehemently by naʔKsɔ? and we can expect individual preference to affect harvesting choices. However, it is true that I have never eaten yams during the fruit season. I have recorded only one exception, which occurred in a camp in Kəciw (1993) where there were no fruits. The longest period when there are no cultural objections or practical obstacles to yam-digging would be from October/ November (when the fruits cease) onwards till May or June. By then, the honey will have arrived and the early fruits appeared; less and less time is spent digging.

Digging still carries high cultural value but the dietary significance of yams has been replaced by rice. Before monetization of the economy, rice was exchanged for forest products. Availability depended greatly on trade relations and accessibility to the traders. Even during the Endicotts' early studies, roads had not yet penetrated the interior of Kelantan and rattan shipments were transported by land and water.[6] During the flood season, people could be cut off. Today in Pahang, rattan-collection is done more consistently, traders can drive up regularly, and money flows more continuously. When rice stocks run out, people can just buy some more. Even when there is rice in camp, the Batek may still go yam-digging but the yams are no longer an absolute dietary need.

For Orang Asli communities living in marginal situations, even in sedentary conditions yams continue to be an important staple. Kuchikura's study population, recently sedentarized in the 1970s, were highly dependent on bought or processed foods for their sustenance, with 70 percent of the total energy intake coming from "exogenous" sources. They actively harvested wild yams when agricultural foods became so scarce that at least one meal a day had to come from the forest and when these foods had run out (for up to a week at a time). The rations given by the JHEOA were neither regular nor dietarily sufficient providing (for example) perhaps only 20 percent as much protein as could be obtained from blowpipe hunting.[7] With the Pahang Batek the dynamics are slightly different: they do not depend on JHEOA for subsistence needs and because they do not farm they are not vulnerable to the fluctuations of agricultural harvests.

In view of its decreased importance, what makes people want to dig? Well, digging for yams is an adventure much like hunting. The same words are used to mark success or failure in hunting and digging: *bərguh* (to succeed), *pawɛs* (to fail), *siyal* (can't find animal tracks or tuber vine), *malaŋ* (can't find animal tracks or tuber vine because someone has broken a foraging taboo). In short, what is said of hunting is said of digging as well. Digging like hunting creates its own context of fun: "the laughter and conversation that mark their digging expeditions suggest that they do get some enjoyment and satisfaction out of the effort."[8] And excitement: there is the *search* for a likely-looking vine; the *anticipation* of digging at the right point of entry and as the blade digs ever deeper; the *tracking* of the root as it travels through the ground; and finally the *satisfaction* of procuring fine-looking tubers.

Within the general context of work, the tendency is to "go to the forest" everyday for a short stretch then relax and rest up, then go back in again. According to na?Gk and na?Abr, their enthusiasm for digging rises and dips depending on how successful they were. If a dig is unsuccessful or there is a pattern of failure, disheartenment sets in and digging expeditions cease. Other women express similar sentiments; they say when a dig is unsuccessful they feel *males* (a general purpose adjective suggesting displeasure or reluctance). Likewise, hunters say they will have a run of success followed by failure and, often, suspension of efforts and switching to alternative activities. Those who have not been hunting for a while will begin to express their longing to get back into the flow of things. Someone else's success is often the signal to strike up one's efforts again, not only to satisfy competitive instincts but because it shows that the stuff *is* out there for the taking. Successes and failures are some of the "micro" (psychological) reasons for this start-and-stop pattern of work. Outside the forest, this pattern creates adjustment problems for *employers.* For example, Razha was told by the Kintak Bong's employers that "the Semang are indolent and that they will only work if there is somebody to pester them. Furthermore they eat too much and at irregular intervals, so that the employers generally refused to employ them anymore."[9] It is ironic, of course, that irregularity is precisely the quality that promotes *conservation.* Suspending work from time to time allows prey populations to adjust and recover. And whether it is intentional or not, the overall effect is to lighten one's traces on the land.

Ecology, Variety, and Preferences

The variety of yam species affords the Batek a great deal of choice. As ?eyGk noted, yams like flat, level forest, with somewhat sandy soils and they are found in both mature and secondary forest. As a body of foodstuffs, then, they are widely available.

However, species distribution is not even. Some are uniformly dispersed: therefore, more widely harvested and generally known. One such species is *D. orbiculata,* which grows throughout disturbed and undisturbed forests (Batek *takop;* commonly harvested by Batek and Semaq Beri, which is tellingly also the name of the plant genus).[10] Old people like ta?Sudep (in his late sixties) remember a past practice of subsisting mainly on *takop* and game. Other species cluster in specific ecological zones: as such, these are encountered and harvested only in special or serendipitous circumstances.

An important consideration is the *lis* (growth habit of the underground stem) of particular species, which determines how much effort it would require to harvest them. While most tubers are buried deep in the ground as protection against rooting pigs and other wild animals, some are found quite close to the surface of the soil; to compensate, the latter protect themselves with poisons, coarse thorny roots, or leaf-irritants.[11] The Batek rank the deeply growing ones among the most difficult to harvest; on the other hand, poisonous tubers have the advantage of ease of harvesting but the disadvantage of requiring much effort in processing them for human consumption.

Work-wise, there is always a trade-off between eating something good that is rare and the energy and time involved in searching for and making them fit for consumption. *Takop* may well be a dependable famine food but it grows not just deeply but extensively over a broad area. *Wɔh*, a regular favorite (it produces fairly thick food-filled tubers) does not grow as deeply as *takop* but it spreads far and wide as well. At the other end is *tampak* whose tubers grow in discrete bunches of just two or three pieces close to the soil surface. Even better from some people's point of view is *rɛm*, which does not travel far from the point of stem emergence and has an enormous tuber: dig just deep enough to get to the tuber and you'll have enough food to sustain the whole camp. Energetically this is one of the most efficient yams to harvest; Kuchikura reports a maximum weight of fifty kg.[12]

Unevenness of harvesting patterns is shown by the difference between my collection of proper yam names, thirty-three, and Endicott's presentation of eighteen names from Kelantan (two for a single species).[13] No two areas will contain identical densities of yam species; therefore, use values, as revealed superficially in nomenclatural knowledge, will also differ from place to place, from group to group, from person to person. For example, Endicott listed *takop, hɔw, wɔh, gadoŋ*, and *rɛm* as the favored species; my list (based on frequency of harvest) would be *takop, wɔh, tahoŋ, pam, kɔnsey*, and *rɛm*. These lists are broadly similar but the differences do point to some heterogeneity in the yam supply. More specifically, Endicott reports that *D. hispida* (Batek *gadoŋ* or *kɔdat*) was relatively abundant where he worked and was commonly harvested.[14] This yam, along with *D. alata* (Batek *ciŋʔil*), *D. pentaphylla* in Kuchikura's study, and *Amorphophallus campanulatus* (Batek *hakay*), is commonly grown in Malay gardens. Among Semaq Beri, *D. hispida* "was found only in the secondary forests regenerated from the Malay clearings" (it alone accounted for 35.4 percent of the total yield there). But the Pahang people never deliberately went to such environments to look for *D. hispida*: it did not occur close enough to justify the tracking and digging effort and, furthermore, they say processing *D. hispida* is too much trouble. It contains the alkaloid dioscorine, a poison that must be leached out before the yam can be eaten.

Tending the Harvest

As shown by the density of some species near Malay clearings, the supply and demography of yams, and their varietal development, are affected by anthropogenic activities. Many apparently wild tubers owe their continued life to the harvesters' care. The word "wild" does not quite capture the complex ecology of yams. The presence of recognized subspecies or varieties is suggestive (see notes for *ciyak*, *tahoŋ,* and *takop* in appendix C). More deliberately, the Batek commonly replant roots (tuber heads). Replanting of tubers is done ever so casually. All that's required is to dig carefully or throw the tuber head back into the ground and bury it with a few scrapes and shovels of the digging stick. This is an example of substantive knowledge that does not need much articulation; there are no spells, no rituals, no incantations, no magical phrases to enliven the proceedings.

According to naʔAlɔr, the urge comes from seeing what a fleshy tuber a particular plant produces; she feels "*sayɛŋ* [love]" for it and wants to propagate it. Other women confirm that they remember those that they have replanted and, most crucially, their locations. After several years, comes the desire to see what has happened to that tuber. In naʔAlɔr's words, they will urge their friends "*ʔuylah! Heh haw təmrɔh* [Hey friend! Let's go to the old digging spot]." Or the digger might say to her friends, "*Wək bab yɛʔ nuy. Yɛm haw yɛm tɔt. Haw sah tɔt leh* [There is my food from before. I want to go see. Let's go look at it]." Over time they develop a pattern of returning to old digging sites and manipulating the productivity of those that demonstrate good potential.

Lexical coding shows that, however informal, random, and unpredictable, these replanting—or propagating—practices are not idiosyncratic; the Batek do monitor the life of individual yams. Nor are the Batek oblivious to the impact of their activities; it is not a simple matter of "dig today, gone tomorrow." In general, there are two types of impacts: when the stems are not killed by the initial harvst, so the yams are preserved for the future (this can be accidental or the result of taking good care in digging) and when the tuber heads are deliberately replanted or propagated.

Any act of digging (for plant or animal foods) is to *bay.* But the verb for "digging" is changed to *pənrəʔ* if the initial dig did not kill the yam and it continues to produce fleshy tubers afterwards; *təmraŋ* if the yam species in question is *rɛm*; and *təntuŋ* if the species is *tahoŋ* or *ciŋʔil.* For replanted yams, *təmrɔh* (mentioned above) specifies that the place of digging contains a replanted tuber. (It is different from *kəpən* [pit or trench] and *tom* [the numerical classifier for each of the holes dug]).

The digging verbs are different again when there has been conscious intervention in the life of the plant. For those yams like *pam* and *wɔh* that regenerate from the mass of roots at the base of the stem (the *bəhiʔ*), the "secondary digging" (following replanting and regrowth) is called *hənul* or *hənwil.* Similarly for those like *takop* that regenerate from the root that connects the tuber to the stem (the *jəliyɛk*), the secondary digging is called *kəmənkah.* Kirk Endicott concluded that "during the study period [1975–1976] people did not make any deliberate efforts to 'manage' the stuber supply, but they may have done so in the past." In my experience, then, this is not a "past" but an ongoing practice, whose implications are still not understood. We will examine more implications of the Batek's tending practices in the discussion of fruits that follows.

Flowers and, Most of All, Fruits

In view of their perceptual, cultural, and ecological salience, fruits and flowers need to be discussed separately. The flowering season (beginning around March or April) precedes the fruiting season (July to October) and overlaps with the honey season; these months are themselves highly salient. This section will examine in some detail the importance and exploitation of forest fruiting and its influences on Batek behavior and consequently their perception of the *həp*.

A Flash of Welcome Color

The visual world under the forest canopy is dominated by the greens and browns of vegetation. In the main, the forest lacks brilliant colors. Flowers bring an all-too-brief flash of welcome color to the forest ambience. Most flowers I have never seen. They may come from fruit-bearing trees, gingers, shrubs, palms, epiphytes, climbers, and so on. Indisputably, flowers are relatively rare in time and space; when there are flowers everywhere, they evoke great joy and moments of playfulness. People will collect some of the flowers, especially the gingers, for bodily ornamentation (some are tabooed from such use). A common scene is of a hunter returning to camp with a flower in his hair, and the flower confiscated within minutes (moments?) by his female relatives or children. I certainly appreciated the arrival of these bright colors in early 1996, after spending weeks on end in the damp and darkened forest. In Taman Negara, the air was festive. Girls and *kəradah* (young unmarried women; plural of *kədah*) ooh-ed and aah-ed over the flowers and often went out collecting them.

A Time of Great Plenty

The fruit season (*masaʔ tahun* or *masaʔ kəbiʔ*) is clearly the highlight of the year: "a time of great plenty and the happiest time of the year."[15] Regarding the season, Caldecott reports for the Peninsula's dipterocarp forests:

> although individual trees or species may flower or and fruit idiosyncratically (and some fruiting continues all year round), a markedly greater proportion of the trees in any forest sample will be in fruit from April to October than at other times. In most years, 3–15% of all mature trees fruit in the peak month, but every few years their number is supplemented dramatically by a so-called "mast" fruiting of dipterocarps.[16]

The peak months in Batek experience would be from July to September. As these months approach, there is a great sense of expectancy. Camp groups disbanding in March or April would have been looking ahead to the current year's fruiting possibilities. They would be considering itineraries and companions: where to go, what route to take, who to go with, who to go to. An implicit aim would be to sort out business before the fruit season arrives. People may express misgivings and anxieties, weighing one set of choices over another as ʔeyPaliy did on this occasion (April 8, 1996): "*Yɛʔ rajin bənɛr yɛʔ ŋɔk kaw* [but] *bel ʔoʔ pəsɛn. Tapiʔ ʔajoh— yɛʔ malɛs. Yɛʔ rajin yɛʔ ŋɔk jəmit. Bah-kəntəʔ pun jaraŋ. Yɛʔ ʔiŋit jəŋrəŋ ʔakal yɛʔ bəlaʔ. Tambah pulaʔ tahun bəw pulaʔ* [I'm really comfortable here but my younger sister has called for me. But I don't know—I don't feel like it. I really want to stay put. And I don't go upriver that much. I'll mull over it on my own. What's more, this will be a big fruit season]." This is a good example of the sorts of dilemmas that people will find themselves in: to stay in a place where the fruit harvest is assuredly good, or to join kin in an area of uncertain possibilities and possibly end up eating no fruits at all.

The start of my dissertation fieldwork coincided with the start of the 1995 fruit season. Here is a selection of topics that consumed everyday conversation then (July and August):

(a) what's ripe
(b) what fruits might be ripe
(c) where one had seen a particular fruit
(d) what stage of ripening that fruit might be in
(e) what fruits are not yet ripe but will be soon, and when that might be
(f) organizing a work party to go collect fruits
(g) encouraging a campmate to join in collecting fruits
(h) where one could move camp in anticipation of fruits in that location
(i) reporting to others where one had seen fruits, in what condition
(j) discussing how best to harvest particular fruit

The Batek's enthusiasm often masks some plain ecological realities: like yams, the fruit trees are not evenly distributed around the forest with some patches being particularly rich and others not; the harvest is uneven from year to year; and species distribution (i.e., the location of personal favorites) can be quite localized. Thus, whether the forests are in good quality or not, there is always an inbuilt anxiety about what to expect from the upcoming season, which is not unlike the anxiety of farmers tending domesticated crops: "Much as he may know about his trees, the [fruit] grower in effect has little control over the yield. He has to wait patiently until his trees come into bearing—and when they do, he has to be prepared for crop failures as well as good years."[17]

Only with probing was I told that the Batek already recognize this annual variation linguistically: a *tahun cəkey* is a "lean fruit season" and a *tahun bəw* is a "season of big fruits; i.e., one of abundance." *Tasek skaliʔ* means that the fruits all ripen together synchronically, i.e., in mast fruiting of dipterocarps, and *juw* is the tail-end of the season when few fruits are left and it's time to gear up for the rainy season. One important thing to know about a species is when fruits are safe to eat. The ripening sequence is: *butik* (bud), *klɛt* (raw), *tuhaʔ* (green; some species edible already), and *tasek* (fully ripe). Many larger fruits like the Artocarpus species will never reach *tasek* stage: they are harvested from *tuhaʔ* onwards. If left on the tree

Table 6.2. Ripening stages of baŋkoŋ (*Artocarpus integer* var. *silvestris Corner*)

Label	Definition / description
butik	bud
ʔec dəkɛɲ	lit. shit of bamboo rat (*Rhizomys sumatrensis*). Flowers have budded
palɔh	fruits have budded. The more general meaning of *palɔh* is "leaf sheath"
kəbiʔ	fruits have come out
ləkɛm kawaw	lit. bird brain. An early ripening stage
ʔayam	no seeds yet. *ʔayam* customarily refers to the fragrant leaves worn as body ornamentation
ʔayam kəbiʔ	spikes have emerged; no seeds yet
miyaŋ maliyeŋ	not defined; probably a reference to the tree's latex. *Miyaŋ* means "skin irritant or rash-causing." *Maliyɛŋ* is probably a word-play on *miyaŋ*
cərahuʔ	seeds can choke one to death, comes after *miyaŋ maliyeŋ* stage
kəroʔ cəcol	lit. back of *cəcol* (Himalayan striped squirrel; *Tamiops macciellandii*). Close to *tuhaʔ*, not yet edible
cəɲciŋ	can eat already
sədaŋ kak	lit. choking. If seeds are eaten raw at this stage, they will stick in the throat, but the seeds can be boiled
makɔʔ kwaŋ	lit. egg of Great Argus Pheasant
bəralat	*tuhaʔ* stage
sədaŋ mamah	lit. chewing, i.e., the seeds are edible

too long they will be eaten up by the animals; animals and humans are in a race to pick off the fruits. As Kayəʔ chuckled once, the fruit season is a time when "the animals are fat and so are we."

An Intricate Filigree of Conundrums

Harvesting fruits, unlike yams, involves delicate scheduling. Each of the four ripening stages may be subdivided into even finer categories, especially with regard to the larger fruits. The most comprehensive list of growth stages in my collection is for baŋkoŋ (wild cempedak or jackfruit; *Artocarpus integer* var. *silvestris* Corner). Its various stages of growth are collected in Table 6.2. Thus far it has not been possible to rationalize the names; not everyone can identify what the correct order should be. I also suspect that mine is a composite list from different geographical areas of Pahang and Kelantan, which would lead to much disagreement about how to name different stages of growth and even over the significance of particular stages.

The fruits' variation in time and space, together with socioeconomic concerns (like a sibling's need for one's company or rattan-collection demands), creates an intricate filigree of conundrums affecting movement decisions. One of the more interesting effects of fruit ecology is that it helps in scheduling. Because fruit ripening stages are staggered, the Batek may be able to use the ripening process of different species to schedule their movements. Certainly they know not only when a fruit is expected to be in which stage, but what other fruits are in that same stage at that time, and where it stands in the ripening sequence. Because fruits are patchily distributed, it is possible that they employ this method for scheduling their movements during the season. For example, if one knows that one fruit is undergoing X stage, then one can decide whether or not it is worth the effort to move to another area to search for Y fruit because it may or may not be in the right stage just yet. There is certainly a recognized sequence to the ripening process, but it is not clear at present to what detail the Batek use this ecological information.

At least once I heard a group complaining that they ended up in a place without fruits and so missed the season entirely. Cases like this would be due to other issues taking priority over the fruits, which could be caused by low expectations of the harvest. The main competition these days is rattan-collection. Traveling to fruits potentially has a countable economic cost. For example, they may move to areas where they will not collect commercially (like Taman Negara, where commercial collection is banned). It *is* possible to combine rattan- with fruit-collection, as the Kəciw group did at the start of the 1995 season. I was impressed then by how abundant everything seemed to be and how casually everyone was doing whatever they felt like doing on any given day. The other side of the story came from the rattan trader who complained that no one was supplying him with much rattan! Accordingly the Batek's income for that period was diminished. However, this is balanced by lower cash outlays because people mainly eat off the forest. Other than store-bought rice, they will be living solely off fruits, game, and fish, thus making for a highly varied and nutritious diet.

Fruit Preferences

Fruits are probably the favorite foods.

> Although the wild yams are the staple of the foraging diet, it is seasonal fruits (*kəbʉʔ tahun*) which, of all foods, dominate the thoughts of the Batek. Their basic understanding of the world and themselves is woven through with fruit lore, and their craving for fruit exercises a powerful influence over their whole way of life. . . . The cycles of the seasons, such as they are, are cycles of fruit to the Batek, and the processes of change in the weather are intimately linked to the state of the fruit supply.[18]

Fruit preferences are not clear; the pattern is that the diet revolves around some key species. Ecologically, these may be the keystone species for the Batek: if they appear the fruit season is considered good and if they do not it is a *cəkey*. Among these species, we have already encountered *baŋkoŋ* but its status is somewhat ambiguous (see following). Kirk Endicott reports that the Mendriq consider this fruit the "ancestor of all fruit" and the first fruit to be eaten by man. He adds that the Batek in Kelantan, however, consider *tawɛs* (*Artocarpus* sp.) the first fruit.[19] Some other familiar names include: *cawas* (Malay *buah keledang*; *Artocarpus lanceifolia*), *tampuy* (Malay *buah tampoi*; *Baccaurea griffithii*) and its various varieties, and a range of Nephelium species (from the domesticate *taŋɔy* [rambutan; *N. lappaceum*] to "minor" species like *ramey, pərajok, klas, kəlkil, təkɔy,* and *paʔic*). The list of recognized fruit and nut species is diverse, and my list is only preliminary.

In general, the bigger the fruit the higher its food content and the more attention it gets. *Baŋkoŋ* is the biggest one of all—Endicott gives a maximum weight of twenty-seven kg[20]—but unlike its cultivated congener *cempedak* (the common jackfruit) the *cəŋkop* (fleshy perianths that surround the seeds) are tasteless. I suspect that it is perceived like a famine food and considered boring in comparison to its flashy friends. NaʔKsɔʔ once described an area as having no fruit even though *baŋkoŋ* did occur there. In the late-ripening stages the perianths can be chewed raw but in general it is the carbohydrate-rich seeds (the *mɛ̃t*) that are boiled in fairly large quantities, thus becoming a major staple that can substitute for rice or wild yams. *Baŋkoŋ* may also fruit twice, with the crop lasting beyond September, and so it is available right through the season. (I would revise Endicott's finding that *baŋkoŋ* is only available from September to December;[21] I have seen it harvested as early as July.) Its long growing season enables people to plan their movements well in advance: one can risk traveling to a food-poor area if one knows that there will be *baŋkoŋ* there even if all other procurement activities fail. The seeds also last a long time and can be stored by burying in the ground. I once encountered *baŋkoŋ* seeds making a surprise appearance in February; ʔeyTaniŋ had dug up his store from the year before. Other people subsequently confirmed the practice.

I have not caught anyone in the act of burying the seeds. The most common pattern is to carry sackfuls of seeds from one camp to another until the store runs out. They can be replenished right through the season. Seed storage would be expedient if people know they will return to the same area or within easy trekking distance of it in the next few months; otherwise, the stores would go to waste. Movement plans are rarely this concrete. The pattern is to move from one tributary system to another as social-ecological conditions change and it may be several years before they return to a former resource zone. My estimate thus far is a duration of two to three years before a camp is inhabited again. Storage is likely to be favored by people whose customary range is quite small, who expect that at some point in the near future, they will be traveling within a two- to three-, perhaps even four-, hour radius of the camp where the seeds are stored, certainly close enough that a hunter could be sent to pick up some stores on his way home. Despite its rarity, the existence of the practice certainly challenges the once-accepted cliché that hunter-gatherers are distinguished by their inability to practice food storage or manage their food supply.[22]

Strategies of Exploitation

There seems to be three overlapping strategies of exploitation: to select campsites that are close to groves and concentrations of known species so that fruit trees are accessible within thirty or sixty minutes of walking from camp; then to go on fruit forays farther afield, which might mean walking durations of two or more hours each way; and opportunistically harvesting whatever fruits are encountered everywhere that one goes. These strategies are not exclusive of one another, as I now explain.

Apropos of the selection of campsites—which is in effect the selection of a tributary system to travel in—there is a certain "gentleman's agreement" to disperse camp groups widely. This ensures everyone has an equal chance of procuring something and, further, alleviates crowding. When I first began my studies, I was intrigued how groups managed to avoid banging into each other. The system when explained to me by the likes of taʔKadɔy and ʔeyKaw sounds almost too simple.

To prevent competition, they say, messages are sent to nearby relatives (such as those within the drainage areas of neighboring tributaries of a major river) to keep everybody informed about future itineraries. Decisions are then made based on where everybody else is going. No group seems to be left out of the information network for long. Another reason for keeping track of others' plans is to schedule reunions: for example, two groups of people living far apart decide that they want to spend all or part of the fruit season together and so they keep sending messages back and forth to synchronize movements and arrange for their respective routes to overlap. I showed a detailed instance of this in the movements of the Ruwiw group in 1996.[23]

Though information can spread rapidly, the process of keeping everybody informed and synchronizing movements would take time so the planning, consulting, and scheduling would have started, as mentioned before, months before the season starts. Even after its commencement, members may join or leave a group, or start new ones, so itineraries might keep changing and unworkable plans abandoned. This is the famous strategic flexibility so well described by Kirk Endicott in his publications.[24]

Once a group is on its way, harvesting of fruits will involve everyone except the sick. The size and composition of harvesting groups are variable, from one or two persons to thirty or so. Groups decide among themselves where they will go today. A lot of discussion goes on in the evenings about this and people often wake up knowing how they're going to spend the day. Of course plans might be abandoned if companions change their minds. All this presumes background knowledge of what fruits occur where, knowledge that is constantly replenished as people return to camp with news of the latest. So harvesting is as always integrated with monitoring, with new information being brought back to camp to update everyone.

Fruit trees are not owned so access is universal, but harvested fruits are (in the sense that no one may take someone else's harvest without permission). When a group is planning to journey to a particularly abundant grove, other households may send representatives like preadolescent children to join that group, to ensure that the family also gets a share.

There are constraining factors. Someone may reserve the harvest (usually for a short time only) by leaving a gash on the tree; the next person who sees the gash will respect what it means. Those who don't are considered to have stolen the fruits. Another constraint is distance. Small children are always a problem; they need to be carried. Mixed harvesting groups—especially those dominated by women—do not like to trek too far from camp.

Occasionally smaller groups of people will journey distances (something beyond two walking hours) to reach the fruits. These are what I call fruit forays. In those places, there is a known history of exploitation but the harvesters are less certain of current crop conditions. So the risk of procuring nothing is greater. I once trekked a punishing distance with naʔKsɔʔ and a group of men and children to reach a promised harvest; when we got there we found that we were too late. Others had "stolen" the fruits from us. So we wound round the hills and opportunistically harvested whatever we could find. Commonly such groups might involve fewer numbers, like a couple of hunters. One singular exponent was the late Tan, already arthritic (he was in his seventies when he died), who would disappear on lone forays for hours at a time and almost always returned with something in his sling. The farthest such journey I have made was in 1998, when the Tɔm Lal (Tɔmiliŋ) group took advantage of my visit that day and the extra boat I brought from Was Yɔŋ (belonging to a Batek friend). About thirty adults and children then scrambled into two boats and we motored an hour upriver to exploit an abandoned Malay orchard. This may not seem like far but the comparable walking distance would have been at least four hours over hilly terrain, which is normally too strenuous for such a mixed group to undertake.

Everyone, everywhere, will be on the lookout for fruits at all times. Harvest might be serendipitous. This strategy is described in the literature as "opportunistic."[25] By this is meant that foragers may move easily into a new economic activity just to see what advantages they might derive from it. In terms of specific activities like fruit-collection, it implies casualness and spontaneous picking of one thing after another from the forest. In general, I see no problem with this cast on things. But it does encourage observers to overlook whatever method might exist, which might be implicit and subtle in their aims and effects.

I suggest that the Batek's fruit-harvesting methods are to some degree predecided before the season even starts. In choosing one harvesting area over another, a group has already identified the main species to be exploited and the likely set of harvesting strategies necessary in that place. Once settled in, difficulties, like the fruits being located farther and farther from camp, are resolved by moving to the next camp (on average every 8.5 days). So the Batek certainly know which are the main species wherever they are and by association what the behavior of fauna and avifauna might be. Information from sight, sound, smell, and so on will be noted. Indicator information in the form of animal and insect sounds and traces would also play a part in planning and scheduling. None of this information is consistently predictable but one must nevertheless be alert to what it means or promises.

What is left to opportunism is finding out on daily forays which particular grove or tree is doing well this year and what other fruits and nuts in the vicinity (say, from palms and rarely fruiting trees) may also be available. Quite a number of species (like *blauwɛn*, a type of mango [*Mangifera* sp.]) don't produce many fruits and only occur in isolation or small numbers so it is not worth launching an expedition just to find them—unless for fun and to add variety to the diet. The most efficient strategy for this group of fruits is opportunism, to take the fruits only when encountered.

Harvesting Methods and Their Effects

Harvesting methods depend on the type of fruit, plant architecture, where the fruit grows, and whether it *gǝl* (falls) when ripe. Durian fruits for example fall naturally and can be collected from the ground. In Waɜ Yɔŋ there are some nearby durian trees; the children would make a game of waiting for the fruits to fall, sitting down patiently to watch the trees or running over when they heard the familiar sounds of falling fruits. (This is a favorite pastime of rural Malaysian children, not just the Batek.) It goes without saying that those plants, like the palms and climbers, that produce fruits close to ground can simply be picked off by anyone, including children. Harvesting *baŋkoŋ* is "no sweat": the fruits grow on leafy twigs projecting off the trunk, relatively close to ground. All that's required is to shinny up and cut off the fruits. Anyone can harvest a *baŋkoŋ* tree. For other kinds, like the standard *tahun* tree (such as Baccaurea species) that produces fruits in the crown on spreading branches, the difficulty of harvest goes up with the size of the trees and the distance between crown and ground. The more mature the tree, the bigger the trunk girth, and the harder to climb. Confidence with heights, dexterity, and all-round physicality are definitely prerequisites for all kinds of climbing.

The crown may be very high up indeed. To get the fruits down to ground, skilled climbing is required. In such cases, it is almost always done by teenage boys and men. Beyond a certain size of tree, not every man will feel confident of his climbing skills. Where the trunk is too broad for climbing, the method is to shinny up smaller trees next to the fruiting one, then to *plibat* (move over) from one bough to another, higher and higher, until the crown of the fruiting tree is reached. If the tree is slightly beyond reach, the climber may secure his tree to the fruit tree using a rattan lasso and thus bring himself closer to the fruits. He will be lopping off those branches that are in his way as he looks for a foothold in the fruit tree. Once he has reached the crown of the tree, he will move around it, lopping off the fruit-laden branches. Those below are waiting for this moment. They will be standing off to one side as the branches fall. Once the climber has cut off all the necessary branches, everyone else will rush over, gorge themselves on the fruits, and collect their shares for transport back to camp.

Among lesser trees like langsat and rambutan and other Nephelium species, the branching may start quite low in the ground; they are eminently climbable. Where branching is farther up but the trunk girth is not so great, anyone can climb up to harvest the fruits. Tahaɲ was just seven years old when he went up such a fruit tree to join his older sister Mɛ̃c, then aged twelve or thirteen. But Tahaɲ lost his nerve quickly and had to be guided down (verbally) while Mɛ̃c efficiently lopped off the fruit branches for the waiting group below (six years later, Tahaɲ still remembered the incident when I tactlessly reminisced about it; clearly his loss of nerve had been more traumatic than it appeared at the time). The oldest fruit-tree climber I saw was ʔeyNɔn, who was a wiry man in his sixties. Women young and old are skilled climbers too. It seems that for convenience they will send the boys and men up if the latter are available; if not, they will often attack the fruit trees on their own.

To sum up, fruit harvesting takes several standard forms: to collect from the ground; to *rac* (pluck off), to *tutoh* (lop off the branches), and finally to *cəh* (fell the trunk). The main choice seems to be between *tutoh* and *cəh*. That is, when a fruit harvest is heard about, those who did not see it happen may ask whether the harvester had *tutoh* or *cəh*. This corresponds to finding out whether the tree will still be available for the next fruit season. It is not clear to me why someone would choose to fell a fruit tree; this has never happened in my presence. Certainly the most mature fruit trees are not felled; for one thing, it's too much hard work. (Mature nonfruiting trees *are* felled, when rainstorms reveal that they will fall over a camp and endanger humans.)

As Rambo has pointed out, Orang Asli effects on the ecology can take several forms: habitat modification, direct selection, dispersal, and domestication.[26] All these except perhaps domestication are involved in fruit harvesting. By lopping off the fruit-tree boughs, the Batek open small gaps in the canopy. They thus enable more sunlight to reach the forest floor, promote plant regrowth, and affect the floral composition of those patches. This is a benign form of habitat modification. Upon harvesting a tree, the fruits as mentioned above are eaten on-site and then brought back to camp. The seeds will be thrown away casually everywhere the harvesters go but most densely in the campsites, where seeds from many different species will be piled up by the lean-tos. Some seeds at least will have a chance of successful germination (aided by the increased access to light in the camps and in the place of harvest); the genetic composition of new growth may well be affected by its proximity to seeds from other varieties and species (for example, through back-crossing). As H. N. Ridley observed, this is a common dispersal effect of Orang Asli predation of fruits.[27] It is at least one reason for the increased density of fruit trees along some trailsides and patches. These then provide extra food for microbes, animals, and insects.

Direct selection comes in the form of fruit preferences. Ultimately, the Batek's harvesting practices act as a selective agent because some key species are more vulnerable to human predation and manipulation. It is unlikely however that the fruits will be exhausted; there are not enough people to eat all the fruits off the trees. In my experience fruit season journeys never seem to follow their logical course; even before all the fruits disappear, a camp group will have disbanded and members moved on to other pursuits. Many fruits are therefore left to animals and will rot on the trees. The Batek's practice of spacing out groups and waiting several years before returning to an area also ensures that old fruit patches have a chance to produce new growth and regenerate. They seem aware of these effects of their fruit harvesting but probably do not domesticate favored fruits.[28]

question of intentionality

Table 6.3. Population statistics of fruit season camps, 1995 and 1996

	All fruit season camps	Kəciw 1995	Trɛɲin & Tabɛn 1995[1]	Ruwiw 1996 (six camps)
Mean	40.36	61.38	26.58	35.43
SD[2]	15.22	5.27	7.67	7.87
Minimum	10	50	10	21
Maximum	67	67	43	57
Total days counted	106	29	26	51

(1) Aggregated data for both camps. The individual camp data are in Table 1.2
(2) Standard Deviation

Fruitscapes

In selecting areas of harvest, the Batek have a range of "fruitscapes" to choose from. Fruitscapes are what I call those fruit-rich areas that camps are located in. They are not named or labeled like the landscape categories discussed in chapter 3. Movement discussions during the fruit season show that these areas are always remembered: not only their location and what they contain but the time lag since someone last lived there. They include groves planted by long-dead ancestors, which can therefore serve as time and territorial markers. All the fruit groves I know about are durian or langsat groves. The most idiosyncratic origin of a durian grove I heard about was that sowed by the grandfather of Səraməh and ʔeyBəgok, somewhere in Taman Negara. Apparently the night the twins were born, the grandfather went off to the forest and planted a grove of durian trees. What motivated him is unknown but it was clearly connected to the birth of the twins (now aged sixty). The grove remains and is thus almost as old as Taman Negara. Such groves are remembered and the Batek may go on forays to seek them out. For example, Səraməh in telling me about them was able to name a few nearby ones in Taman Negara: Tɔm Set, Tɔm Bahalak off the Tahaɲ, and along the Tɔm Tənər (Malay Tenor).

In the literature, the most consistently documented fruitscapes are those groves and orchards planted by Malays. One reason why these orchards become available to the Batek is that the Malays tend to move out from the forest interior: in Lipis district, they move closer to the main trunk road; in Kelantan, they were resettled from the Aring to protect them from Communist insurgents. In the 1970s, Kirk Endicott observed that the domesticated species (chiefly various varieties of durian, rambutan, and langsat) found in the Malay orchards were "the most important."[29] The Malay villagers had left behind so many trees on the Aring that large numbers of Batek would congregate to eat off them during the season. Some of my Pahang friends participated in and remember these Kelantan fruit season groupings. In Pahang proper, there is no evidence of similar events. The Batek display a distinct bias for the seasonal *tahun* (i.e., those that are not cultivated) and in all the fruit travels I have documented they select areas where these are naturally abundant. Furthermore, the Kelantan domesticates were abundant enough to support large populations. My data (Table 6.3) on the other hand shows that fruit season camp populations tend to hover around forty, which is close to the general average of 36.2 (see Table 1.2). As pointed out with regards to yams, botanical complexes are heterogenous from one area to another. This difference in fruit-harvesting strategies is another instance of that heterogeneity.

In essence, the most common fruitscapes in my experience are those dominated by the *tahun*. From one area to another the species profile is different and usually includes some mix of domesticates and nondomesticates. In the Ruwiw camps, for example, the harvesters also sought out the domesticates rambutan and langsat as well as what appear to be wild varieties of mangos (*Mangifera* sp.), including one that was so sweet and acidic that it could not be eaten in large quantities and was used mainly as fish poison. There were no durians, a staple of the previous year's season in Kəciw, or the many wild varieties of rambutan that occurred in Treŋin the previous year. I am convinced that, though diverse and extensive, it was a localized harvest. We did not traverse beyond the Ruwiw system, parts of which have a heavy limestone component. Even now, there are many more fruits that I have never encountered; they have not occurred in the areas where I traveled. These geographical differences ensure that when the Batek are planning their fruit season travels, personal preferences for particular species of fruits will have a chance of taking priority over brute gastronomical needs (for example, in a time and place where fruits are scarce, just eating fruits would be all that one could hope for, let alone eating only fruits that one likes).

How Wild the Forest?

This chapter has presented—albeit anecdotally—examples of groves and orchards and suggestions that the Batek's harvesting and eating practices have genetic implications for the forest. Although the Batek do not selectively domesticate and breed plant and animal species, they do have a series of informal and unsystematic practices that—to borrow a term from Douglas Yen—"domesticate the environment."[30] We have seen the Batek monitor and intervene in the life process of favored foods, return to harvest upon regrowth and regeneration, and track all-round availability of supplies by tracking each other's movements through the landscape. They do not practice agriculture and yet they transform the environment, improving its productivity on their behalf, and "looking backwards" to evaluate the effects of their predation on biological populations. Ironically, this is the environment that outsiders coming in after them consider as a "pristine wilderness."

Thus do Batek practices throw yet another ethnographic challenge to those who would assert a rigid separation between "wild" and "cultivated," as though these are immutable states of being rather than dynamic, transitional processes. This has been the subject of a long anthropological and archaeological debate, which parallels a taxonomic debate concerning the distinction between farmers and hunters and the companion question of whether the rainforest provides enough food for people to survive without recourse to agricultural products.[31] It is not necessary to review these debates here. Contrary to external perceptions, Batek do realize that their behavior has effects on the land. We were introduced to this idea in chapter 4's discussion about population growth concerns (pp. 82, 87–89); now we have also seen this consciousness in yam and fruit predation.

Distinctions between "wild" and "cultivated" (reminiscent of the Malay distinction between "wild" and "tame") are useful heuristics but they may be exaggerated. In Christensen's recent ethnobotany of Sarawak's Kelabit and Iban farmers, she makes a distinction between *cultivated* (species that are tended throughout their life) and *semi-managed* (species receiving only little attention and tending, most often only in connection with their establishment and early growth). Following this model, Batek fruits and yams would belong in the "semi-managed" category. Even to this trained botanist's eye, however, "The distinction is not always clear-cut. It may be difficult to distinguish whether plants are semi-managed or not if they are found in a more wild environment, i.e. secondary forest; a seemingly wild-growing fruit tree may actually have been sown and tended the first years by somebody's grandfather years back."[32]

It is possible that many ecological studies of "seemingly wild-growing" fruit trees in the Batek's (and other Orang Asli's) forests have not sufficiently considered this problem. Regrettably, a study of the spatial distribution of fruit groves, as well as their effects on animal populations, has not yet been done, so we cannot assess the degree to which the Batek's contribution to the forest mosaic also supports the reproduction of the wildlife. The broader implication concerns the definition of forest: if so much of the Batek landscape has felt the effect of their tending and sowing practices—including the fauna that depend on these species—then to what extent can we section off the "wild" from the managed? It is for this kind of reason that we must question any presentation of the forest as "wilderness."

This issue is related to a broader bias against anthropogenic forests. The bias is institutional and systematic. De Jong et al., reviewing the case for giving renewed attention to tropical secondary forests, point to the obstacles involved: "the forestry sector in tropical countries has a tradition of focussing on primary forests, secondary forests are often considered waste land, their property right status is often more ambiguous than that of primary forests, and there is little knowledge of how many of the goods and services that primary forests provide can be substituted by secondary forests."[33] In short, scientific attention has traditionally placed its focus on primary forests, on valuing the "natural" and devaluing the anthropogenic. The idea of forest reserves, that valuable timber species should be sectioned off and let grow in their "natural" ecological context, is an example of this kind of bias; it overlooks the potential role of traditional users in altering, maintaining, and even enhancing the biodiversity. As Meilleur has pointed out, not all human activities are destructive. There may be "keystone societies" whose practices enhance rather than impoverish biodiversity.[34] One cognitive effect of this bias for the pristine, as noted by Christensen above, is that scientists are poorly placed to acknowledge ambiguities in the forest landscape, especially when those ambiguities are the result of human modification and selection.[35] It is no accident, then, that comparatively so little scientific credence has been given to the goods and services

I'm not so much making points as I'm showing the time & space of this discourse → the concept of anguish by myself

provided by clearly modified landscapes like fallowing swiddens. But the bias is also political. The distinctions between "wild" and "domesticated," and between primary and secondary forests, are institutionally entrenched; the latter is commonly designated as "statelands" or "production forest" appropriate for timber extraction and land clearance. My position is that even secondary forests are important for conservation and primary forests are not as pristine as imagined. Both types are important to local peoples and serve different roles in social reproduction.

As I said, this chapter is about loss. Or, rather, it is about the negation of loss. In chapter 4, we entered the cosmological realm to see how this landscape was created for the Batek to use. Taking a more empirical approach, this chapter has examined how they use the landscape and promote the regeneration of its resources. We have looked at several social-spatial contexts of use and therefore of knowledge reproduction. To bring all this back to Tebu's message, there is anxiety that fruits are less available now than before. The Batek relate all this to climactic upsets like increased rainfall due to forest clearance activities. What are the impacts on the Batek? The dietary and nutritional implications seem evident. Cosmologically, as Kirk Endicott explained, the production of fruits is intricately connected to the religion.[36] As fruit harvests decline, the forest looks increasingly vulnerable and its reproduction beyond the control of people. Ultimately there is a sense that the forest is becoming more and more unstable and its reproductive cycles less predictable from one year to the next.

Notes

1. Endicott and Bellwood 1991; Foo 1972, 50; Kuchikura 1987, 47; but see Dentan 1991.
2. Kuchikura 1987, 47.
3. R. E. Holttum, cited in Kuchikura 1987, 48.
4. Kuchikura 1987, 51.
5. Karen Endicott 1979.
6. Karen Endicott 1979, 49–50.
7. Kuchikura 1987, 49; 1996, 148, 166 (Table 9).
8. Endicott and Bellwood 1991, 167.
9. Razha 1973, 32.
10. Endicott and Bellwood 1991, 168.
11. Kuchikura 1987, 48–49.
12. Kuchikura 1987, 50.
13. See Endicott and Bellwood 1991, 161 (Table IV). In his dissertation (Endicott 1974, 34) he mentions many more species names than the eighteen formally tabulated in this later study—including a few that appear in my vocabulary—however those species were never botanically identified.
14. Kirk Endicott 1974, 41.
15. Kirk Endicott 1984, 37.
16. Caldecott 1986, 53.
17. Verheij and Coronel 1992, 17.
18. Kirk Endicott 1974, 43–44.

19. Kirk Endicott 1979a, 51, 55.
20. Endicott and Bellwood 1991, 164.
21. Endicott and Bellwood 1991, 165 (Table VI).
22. For one proponent of this position, see Testart 1982.
23. Lye 1997, 258–97.
24. For example, Kirk Endicott 1984.
25. See for example Benjamin 1973; Brosius 1991; Dentan 1991.
26. Rambo 1979, 60–63.
27. Ridley 1893.
28. Endicott 1974, 46.
29. Endicott 1974, 46.
30. Yen 1989.
31. For a review of domestication theories and concepts, see Harris 1996.
32. Christensen 2002, 50.
33. de Jong et al. 2001, 564.
34. Meilleur 1994.
35. See also Dove 1992, 246–47.
36. Endicott 1979a.

7

To See, to Hear, to Walk, and to Know

To the rank outsider, all that can be seen of the forest is the topography and vegetation, its *materiality*. I was frustrated and discouraged never to see the forest for the trees. I could understand the forest theoretically, but not *see*, or know intuitively, what tied all of "it" together. I longed, desperately wanted, to get "into" the forest, to know it like a native would; but as months went by I began to suspect— this will never happen. Like the colonial officials that briefly appeared in chapter 5 (p. 95), I had moments of hatred for what I was doing in the forest and rejection of everything that it—and the Batek—represents. Those officials, we'll recall, felt oppressed and in their lowest moments saw nothing in the forest. But that's not the only possible reaction. The old Malay villagers didn't see *nothing*. They may have feared the forest but that's because they saw spirit beings and believed that these immanent forces could be disturbed by human presence.

I've made the contrast between having and not having a view (pp. 95–96). I don't mean to imply that the Batek see the forest the right way and we others don't, that not having a view is the proper way to see. I find views pleasurable. Looking down from a mountain peak, or any comparable high-up vantage point, there is no visual anchor. The eyes can dart from one point of interest to another. True freedom—no social inhibitions—an exercise of power. To have a view is to command a position *over* the landscape. In the early months of my dissertation fieldwork, I often fantasized that I'd rent an aircraft to obtain a bird's-eye view of the forest! After weeks of traveling on those lowlying meandering pathways, I thought this was the only way to get the geography of the place straight in my head—rise above it. But what we get in visual range from the top, we lose in detail, sensory engagement, and bodily experience.[1] For what is viewing: *capture* the landscape, freeze the flow of images that the intellect may grasp what is going on. Yet to know the landscape like a native we *must* attend to the flow, and this means surrendering the view. Even then, I came to understand, "native knowledge" would always elude me. Somehow, I learnt not to mind.

Though I was born and bred in Malaysia, I was twenty-two before I even stepped into the rainforest, and then as a tourist looking for adventure. Under Batek tutelage, beginning some five years later, my perceptions began to change. Initially, it was the memory of that romantic first trip that led me to consider doing fieldwork in the forest. Once I was "in," the forest just happened to be the place where the Batek live. I prided myself on my pragmatism; no romance here, thank you!

Stepping down from a rise onto a flooded hollow on August 20 1995. Early morning; still misty; moisture everywhere. There was a sort of sheen over everything. Usually I hated getting my feet wet—shoes took ages to dry. Kayɔ?'s scolding rang in my ears: "Get a little bit wet, never mind!" On this morning, as always I grimaced as I poked my foot at the water. And suddenly I liked the feeling. I was wading. I was sloshing water around. I sloshed extra hard just to show how well I could do it. I was sorry to reach dry land again.

I think what happened was this. In the forest I had no linguistic, cultural, or geographical bearings; I couldn't make sense of my environment. I was a child, I was like a child, I was expected to be a child in learning every task as a child does. The fully formed adult in me rebelled against this. Wading through the water reminded me—though viscerally—of many childhood moments, a bit like the reverse of cognitive dissonance. I had grown so used to hating this sensation but suddenly it had become as pleasurable as capturing a mountain-top view. Such counterintuitive moments became less and less unusual and then I stopped noticing them altogether. Something in my perceptions shifted, so subtly that I still can't explain it: like the two halves of my life, "before" and "after" the forest, had momentarily assumed their rightful unity. I minded the Batek's lectures and scoldings less and less. I appreciated more the position and role of the child, relearning everything. (Of course, as the child "matured," so did the adults around me lessen their carping.) In line with my overall pragmatism, this meant I was coming to accept the forest—wet, dry, whatever—and to resign myself to never knowing it like the Batek do. Short of "going native," the child in me would never fully mature. And I appreciated the value of this: standing just on the cusp of the group, neither in nor out, able to look in and look out at the same time.

We go to the forest bearing with us socially shared images, ideas, and knowledges of the forest. How we see is closely related to what we know, or think we know, about how things should be. In our arrogance, we want to take a position in the landscape: to poise ourselves to view. Humiliation is the reward when we cannot relinquish that position. A formidable obstacle is our customary sense of space and appropriate land use: habits of perception shaped by the representations we are conditioned or trained to value. Lest we fall into the trap of relativism, some kinds of seeing are clearly dominant, have political ascendance, and the power to bring change into focus. Representations are not just artifacts of culture but, as with the marginalization of mobile peoples (pp. 96–97), serve political ends. All this parallels the tendency to privilege certain kinds of knowledge, such as that based on empiricist, rationalist thought. Closely related is the value accorded to certain kinds of actions over others. In the Peninsula, the formal term for forest reserves is Virgin Jungle Reserves (VJR): not only is the desired stand a *jungle*, it must be *virgin*; i.e., twice removed from human interaction. When we go to the forest carrying a certain mindset about it, we reproduce such representations and all their political and economic implications—in the case of forest reserves, the exclusion of local communities so that timber reserves may be protected.

Objects of perception do not just present themselves for spectating either. They may have been created, named, shaped by historical forces. In Southeast Asia, ricefields, whether of Malays, Ifugao, Balinese, or Javanese, exemplify a type of landscape that's been sculpted by generations of social labor. These have strong aesthetic power. While to some they may represent culturally modified landscapes, to others they may represent appropriate forms of nature. Landscapes have ambiguous values. To recall the previous chapter's discussion, even a seemingly wild forest may have felt the impress of human agency. If we go to the forest expecting to see virgin jungles we may find them, but never really *see* the forest. And so the best way to enter the forest really is as a child.

Whenever some kinds of seeing are socially approved, others are *denied*. As often pointed out, it's remarkable how little the average non-Asli Malaysian knows about the Orang Asli.[2] I myself grew up not too far from Semai homelands in Kinta Valley but I didn't realize that until I had left home. Until, that is, I had obtained the distance necessary for a second look. I had not been socialized to *see* Orang Asli. But the landscape changed. Now you may see Semai settlements and agroforests from the new North-South highway. This is spectating. Orang Asli lands have become specks in the landscape that others travel through: peripherally in the line of vision as the vehicle speeds away. Orang Asli are now in view but, arguably, still "out there," exotic, distanced, and unknowable.

All of this is a prelude to the point of this chapter, to move in for a closer look at the forest and how the Batek know it. Theirs probably ranks among the most peripheralized visions of all. Our point of departure was chapter 3. That was my attempt to mimic the view from the mountain-top; the overall objective picture that one can get when the landscape is examined in its entirety. We lost many details, of course. The gaps were filled in incrementally in the preceding three chapters, fetching up with a discussion of the social labor that goes into the manipulation and reproduction of important foodstuffs. Now the basic ideas, reflections, and practices are behind us. The first part of this chapter outlines the physical basis of perception: the visual constraints of the forest that have bedeviled so many visitors. Then, arguing for the significance of sounds, the auditory environment, as a compensation for these constraints, I present some ways that the Batek hear, interpret, control, manipulate, and even fear sounds. The sounds discussed in this section are to some degree culturally organized and knowledge of them demands a finely honed sensibility of, and sensitivity to, the nuances of a place. Finally, we stand back a little and examine the social-spatial environment that makes all of this possible. The conceptual objective is to draw attention to different ways of knowing the forest and, reciprocally, the forest's role in reproducing knowledge and therefore enabling social continuity.

Seeing in the Forest

Under closed canopy forests, a high percentage of the sunlight is cut off before reaching the forest floor. Adding to the dimmess is the "obstructiveness" of vegetation. Having the ability to pick out salient details from the mass of wood and green matter is essential in this kind of environment. In foraging, one has to pick out tracks and traces of prey and plant foods in rather dim light. Of moving targets (for example, birds in flight; squirrels darting along a tree limb; swinging gibbons), all that is visible may be a quick flash of moving color, the tip of a tail, an indistinct part of body, or, worse, the shuffling of leaves or rocking of branches. Size, shape, and distinctive markings cannot be reliably determined from ground-level.[3] Time of day is also relevant. In the daytime (*kəlpah*), little of the famed faunal diversity, other than inedible insects, bloodsucking leeches, and other invertebrates, can be seen.[4] Birds are most active at dawn (*mɛ̃t kəmay* or *gəjəm*) or dusk (*ripat*) when visibility is reduced even more. Only in the pitch blackness of night (*haɲɛp*) will many animals come awake and move about—like the tiger, featured in chapter 5. Sometimes a civet cat will be moving through the canopy at night and if it comes close to a campsite, a flank of blowpipes will be waiting for it below.

There are two main implications. First, and simply put, the Batek need sharp eyes. They need to glimpse out subtle details of shape, size, movement, and color shifts under conditions of low visibility and intimate spaces (no more than tens of meters in any direction) where there is never truly a broad vista or view. This skill is evident every time one goes night-hunting (especially for frogs) with the Batek. However, while the Batek are arguably visually adapted to the environment they live in, seeing is something else altogether. This is a skill honed over many hours and years of practice. In my early optimism, I thought I could be taught verbally how to do this. I was learning to dig for wild tubers with na?Abr. She ordered me to look for (*lawak*) the climbing stems of the underground tubers. I asked, "how?"— a rather odd question in retrospect. And her answer, which to me offered no guidance, was: "Just look for it. It will be entwined around the tree trunks." What she did not—and probably could not—say was *how* to look, and how to tell which of the many vines entwined around the trees would be the correct ones. This is a very basic form of vernacular or knowing-how knowledge, which can only be learned "on the job," through trial and error and practical experience.

The second major implication is that by far the greatest fund of environmental information must come from sounds. Visibility is already truncated by the dimness of the forest; how much more so if one is surrounded by standing trees blocking one's vision everywhere that one goes. Obviously, sounds travel farther than visual images and can convey a lot more information: of identity, location, direction, distance, and time. This reliance on sounds has been described for other forest peoples. For Mbuti of the Ituri Forest, Ichikawa found that they "often could not identify the captured birds by their figure alone, but immediately identified them when the birds emitted their peculiar calls." Gell's comment that Umeda on their forest treks tended to use their "ever-receptive ears" to survey the far-off while keeping their eyes focused on the nearby is—with modification—a fitting description of Batek habits too, as we'll find below.⁵ Given the importance of hearing, let us now examine in some detail a few ways that the Batek know the environment through interpreting and manipulating sounds.

Hearing the Forest

A relatively understudied dimension of the forest is the communication system within it, the flow of auditory information from one entity to another. My own data remains preliminary. Nevertheless, I present some observations and impressions here, to suggest how the Batek order the sounds they hear. With the visual field in the forest so impoverished, the compensation is to be hyper-alert to sound shifts and changes and to know what they suggest or imply.

As far as I know, there is not an extensive vocabulary for describing "hearing"; the one word used for all occasions is *kəjeŋ* (from which, aptly enough, is derived the verb for "to notice," *keŋjeŋ*). Under the broad category of *kəliŋ*, which can mean "sound," "voice," or—at a stretch—"language," there is certainly a whole universe of natural and man-made sounds, which unfortunately has eluded comprehensive documentation so far.

Birds and Mimics

Let's start with birds (*kawaw*). They are the most obvious nonhuman communicators of auditory information. They can cross distances and ecosystems, bringing information from one domain to another: from river to land and back again, from treetop to ground, from place to place. And while their "arrivals" in any place can be expected or determined by carefully attending to the "sound universe,"⁶ their points of origin (especially for seasonal or migratory birds) will always to some extent remain mysterious. Perched or flying high, they see a lot more; they have the "view" that is denied to the humans down below. And they have distinctive "voices" that carry far as well as certain habitat- and food-preferences. These general characteristics render them highly suitable channels of "preemptive," often mystical, information, i.e., warnings, indicators, predictions, omens, and the like. One can hardly pick up any detailed ethnography of forest peoples without encountering some mention or examination of the local ornithology, whether from the environmental, linguistic, or symbolic dimension.

Birds have universal appeal. Of course, birds may fall in the category of "pests" and provoke hatred rather than admiration. But admiration is a common theme in popular culture. Birds excite, incite, inspire.[7] The Guatemalan quetzal, the American eagle, and the Papuan bird of paradise are a few examples of birds that have important positions in national iconographies. Popular culture is an important catalyst of emotions: J. K. Rowling's phenomenal success with Harry Potter has given owls—once so feared—a gentler, warmer glow in recent years (in the Harry Potter novels, postally-challenged wizards use owls to send mail to each other).[8] If the popularity of birds is still in doubt, Birdlife International (self-described as "a global partnership of non-governmental conservation organisations with a focus on birds") claims over 2.5 million members worldwide.

For forest peoples like the Batek, apparently there are two outstanding characteristics: birds are difficult to spot in dense foliage but auditorily prominent. Small wonder that the names of birds will often be the onomatopoeia of the sounds. When the people hear the bird-sound, immediately they'll know its name and correspondingly the sight of the bird is not important for identification. Among Tzeltal of central America, nearly half of the bird names were derived from the onomatope of their calls. Among Mbuti (already mentioned), onomatopoeia is even more marked, with calls and flight sounds being the basis for 67 percent of the etymologically traced bird-names. As for Penan Benalui of Kalimantan, apparently mammal names are rarely onomatopoeiac, but many bird names are.[9] I do not have similar statistics for the Batek but I suspect that, like the Penan, they reserve onomatopoeia for birds, insects, and amphibians, rather than the more easily spotted mammals. Examples of Batek bird-sound onomatopoeia: the yellow-crowned barbet *toʔrɔ̃w* (*Megalaima henricli*; call: *toʔ-toʔ-toʔ-toʔ-rɔ̃w*); the Greater Coucal *didit* (*Centropus sinensis*; call: *di-di-di-dit*); the Malaysian Eared Nightjar *captibaw* (*Eurostopodus temminckii*; call: *cap-ti-baw*); and the Indian cuckoo *pumpakoh* (*Cuculus micropterus*; call: *pum-pum-pa-koh*).

As I've mentioned, birds have a firm place in the Batek's everyday consciousness (p. 61; see also Figure 3.1). The indicator birds ("predictors" in Ichikawa's terminology[10]) are probably the most interesting for the present discussion; these birds communicate or signal (*ləʔ*) the onset of certain times and seasons or the future arrival of various categories of human and nonhuman visitors. For example, the *saɲɛt* bird's appearance is supposed to foretell thunder; a number of other birds are associated with the afternoon (*bərəy*) and the onset of the flood and fruit seasons. Research on this issue is only just beginning but it appears that birds are more likely to be indicators than other animals. Like the Mbuti, the Batek have recognized, and encoded in myth and belief, close associations between important birds, species of fruits, and a range of (in)vertebrates.[11] Not all such associations point to indicator relationships; some may just be commentaries on similar morphology or "accidents" of ritual and belief. But quite a few should certainly be regarded as heuristics for further ecological study. The value of this form of vernacular knowledge is that long-term residence in the same ecosystem allows local peoples to detect ongoing patterns of association, which to outsiders may only appear as isolated conjunctions of time and place.

As an example of indicator relationships, consider the Indian Cuckoo *pumpakoh*, whose origin myth was discussed in the context of *haʔip* sentiments

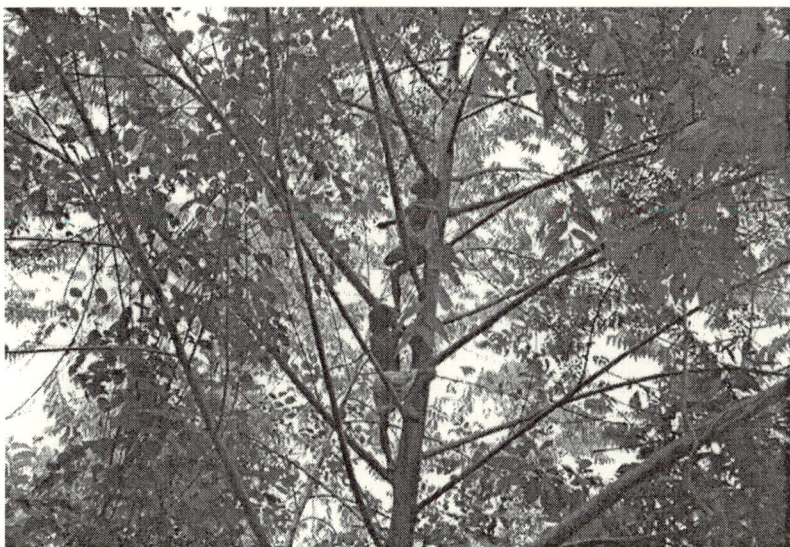

Plate 12. Birds are the favorite target prey of children. The nine-year-olds in this photograph are practicing what is called *tarɛn*—hunting from the treetop. Yuŋ-ʔatoʔ, August 1995.

(p. 100). It *ləʔ baŋkoŋ* (signals the *baŋkoŋ*) meaning that if it calls at the appropriate time of the year *baŋkoŋ* fruit will come soon. It is also known more generally as a "fruit season bird" perhaps because it is most commonly heard around these months, July to October. However, it is also an indicator of the honey season which comes earlier, May to June. The owl *kwoŋkwɔc*, known widely as a "ghost bird," is an indicator of both the tiger and, following its origin myth, *tampuy* fruit (*Baccaurea griffithii*). Children are wont to mimic the owl's hoots using whistles made from half-rinds of *tampuy* fruit. According to adults like yaʔKaw, doing this in the afternoons will call the owl to them; it is therefore dangerous. The Batek, like the Temiar,[12] believe that the owl rides on the back of the tiger: as taʔKadəy describes it, the owl *ʔoʔ kipəy ʔoʔ bat* (it flies around, it perches [on the tiger]). So if the owl is attracted by children's whistles and flies in the daytime, the tiger will follow. The explanation is part-mythical; what the ecological reason might be (other than the temporal conjunction between owl sightings and *tampuy* fruit appearances) remains speculative for now. To me, this is one instance of a broader complex of taboos constraining unwarranted mimicry: tampering with the communication system in the forest. Some wildlife are too sensitive to ambient sounds, too dangerous to humans, or too humanlike to be mimicked.

Mimicry, the human reproduction of animal vocalization, does have its uses—under properly controlled conditions. There is mimicry of other kinds, like the mimicry of other people's voices, which is considered a form of *tinak* (ridicule). One way to look at the relationship of hunter and prey is to see it as a dance around the control of sound.[13] Animals, most notably some species of birds, also can mimic (*hinɛ̃k*) the sounds and songs of other animals and many birds also have a range of songs. One such bird has a broad vocal range and highly creative repertoire; the only way to tell what the bird is saying, or which bird it is, ta?Kadɔy explained, is to listen to the tune carefully. Among the reasons why birds mimic other birds, camouflage (i.e., hiding one's identity) could be one; the Batek call it deception (*ɲo?* [to lie or deceive]). For hunters, sound mimicry is a useful technique to have.[14] To mimic birdsong by whistling, *pichɔc*, is a common tactic of entrapment. The youngest exponent I saw was the six-year-old Tahat in 1996; given his practiced ease, I'm tempted to think that he was already a veteran.

For adult hunters, if the prey (usually arboreal game) is beyond the range of the blowpipe dart, the stalking hunter can trick it by mimicking its sound. Different hunters might improvise with different methods, depending on the nature of the prey (whistling, drumming on the quiver, etc.). The unwary animal, thinking a friend is beckoning, descends to greet the source of the sound and therefore enters the range of the blowpipe dart. The importance of sound-making and -identifying to hunting is the reason, I suspect, that men are better identifiers of sound than women, who rely more on sight and touch in their hunting (of ground-dwelling animals) and digging activities. Often, whenever I asked a group of people to identify a sound in the air, invariably the women would turn to the men for the definitive answer.

Disruptions and Disguises

The Batek are not the only inhabitants listening in the forest. Within the communication system of the forest, humans can be disruptive. To minimize risk and enhance foraging success, the Batek must control noise emission. They are most vulnerable to surprise and danger in the forest away from the safety of camp. One could encounter dangerous predators or unfriendly outsiders, or be surprised by falling trees or branches and unexpected game encounters. So they tell the children not to caper too noisily. If we cannot hear, we cannot anticipate what is coming next. Those parts of the national park close to the headquarters are the noisiest. There hunting is very poor; with too many people walking about and boats zipping up and down the river at all hours of the day, the animals are *jaŋɛl* (wary) and therefore elusive.

Many a conversation in camp will be stilled when incongruous noises afar are heard; there is dead silence as people track the progress of the sound and try to identify it. In the daytime, human visitors are the most common noise-makers in this category. On the logging roads, vehicles are a common source of noise, especially those of traders who drive in much farther than the average villager or logger. As the hum of the distant truck or four-wheel drive (SUV) approaches, the Batek will listen carefully to the sounds of the engine and try to detect whose vehicle that is.

The chances of being tricked by sound, and the value of trickery, are commonly acknowledged and not only in hunting-and-gathering activities. There are some examples from expressive culture. A folktale I collected in 1993 stresses the point.[15] In this case, a hunter is alone in the forest when he hears what he thinks is the voice of the big-bellied Asian horned toad (*kəŋkiŋ*) inviting him to eat at its home. Unfortunately, the voice was that of an old cannibal disguising herself in the persona of the toad; the trusting hunter met his end, appropriately enough, in the cannibal's mouth.

Examples like this one play with the common motif of the mistaken (acoustic) identity. They betray recognition that we cannot trust what we *think* we hear, and it is always possible to fall prey to trickery and danger. They also illustrate my broader point in this section, that there are multiple sound systems recognized in the forest, whether these be from human, nonhuman, or alien emitters. Deceit and trickery (including sound imitation) is one way of "capturing" sounds for one's advantage, in a "game" in which to mishear may well be fatal.

Enjoyable and Fearful Acoustics

This discussion has necessarily taken a highly pragmatic turn. I have tried to place forest sounds in the context of communication: from one person to another, from wildlife to humans and from humans to wildlife, etc. But aesthetics are important too. The Batek enjoy forest acoustics and are musically, linguistically, and narratively inspired by them. For example, giving "sound effects" to actions, moods, or natural phenomena is a common way to improve the telling of a story. In the discussion of *ha?ip*, we've also found that sounds can be powerfully moving and evocative.

There is, however, a limit to how much one can enjoy sounds. A prime example is the sound of Gubar, the thunder-god. In chapter 2, I discussed Tebu's perception of thunder and rainfall. He suggested that thunder can be a good indicator of environmental change. Disruptive noises from this-world can disturb the other-world; overly loud noises like the *booms* of explosives disturb Gubar's rest and he sends down more rain in his anger. However, Gubar has a double persona (he is also a Trickster-like figure). As such, thunder provokes both fear and fun. On the one hand, it is followed by strong winds, treefall, or rainfall, all of which bring danger to humans and can ultimately lead to land subsidence. Thus there is a complex of taboos, *lawac*, that seems designed to minimize human ability to anger the thunder-god (pp. 61–62). On the other hand, as with other Orang Asli, the Batek find Gubar (here less a deity than a man) hilarious, for what personage would make such deep rumbling sounds?[16] What could possibly motivate those inelegant noises? Both Endicott and I have collected a set of tales that mercilessly mocks the Gubar character. The telling of the tales will often provoke someone in the vicinity to collapse in helpless giggles.[17]

Noise Pollution

There are certainly cultural values about sounds and noises. Sounds are not only informative, deceptive, pleasing, inspiring, provocative, fearful, and funny, they can be annoying.

A major complaint of the Batek about large camp groups (i.e., those exceeding forty members), and the reason why many shun joining such groups, is the volume of noise (*?iŋal*) in them. If one wishes to be close to the people in that camp, but not to their *noise*, a common strategy is to set up a smaller satellite camp less than an hour away. Or members will cluster their lean-tos in widely dispersed pockets around the same campsite, which allows the interspersing vegetation to muffle the noises. Many Batek say they prefer the state of *haŋɨw* (peaceful silence). Sometimes people will be seen leaving the campsite area in the daytime, just to find a place to enjoy moments of *haŋɨw*.

People often say they can't think (and therefore answer interview questions) when children are noisy. The following writeup from my fieldnotes (dated December 1995) shows what children's noises can do:

> This was during the early part of the fruit season and the shamans were holding their singing sessions. One evening at dusk, I was sitting with some of the men outside ta?Sudep's *hayã?* [lean-to]. We heard all kinds of caterwauling coming from the farther side of the *pinlodn* [singing] platform. Which we figured was someone singing. Finally, the noise had grown too loud and we couldn't concentrate on our talk. Sia with some exasperation switched on his torch and threw a beam in that direction. There they were: kids like Kərliŋ and Ke?ep, Gk, Alɔr, Hawaŋ, etc., all dancing round and round in a circle and singing their hearts out.

Interestingly, noises not only foretell or mask danger, they can also be dangerous in their own right. Blasts of dynamite (pp. 27–28) fall in this category. The volume of dangerous noises in the towns, even small ones like Kuala Lipis, is said to be one reason deterring people from lingering there. As I can attest, it is nigh-impossible to "switch off" when one leaves the forest. How well I remember the way urban noises overwhelmed my ears whenever I took a brief respite from the field—the bangs, roars, screeches, whoops, howls, and shrieks of trains, buses, cars, lorries, machines, and people talking loudly at each other. In the forest, I had polished the habit of listening hard (foremost to the Batek's language) and of training my ears to pick out the salient noises in the air. Since I could not tell in advance what was and was not salient, I trained myself not to "switch off" in case I missed something important. So whenever I stepped into the urban ambience, my hyper-tuned hearing would still be "on." I heard *more* than I wanted to and every noise seemed so much more amplified and threatening. Also, my sensibilities had become familiar with the organization of forest noises. Urban noises violate that sense of order since they come at the ears from every which way; I was exceedingly distressed. When I returned to Hawai'i to write up my dissertation, I found even the winds to blow too loud. Once I was left shaking by the sudden onrush of a common lawn mower. I can certainly appreciate why the Batek, who are so practiced at hearing, interpreting, and creating stories and music around sound shifts and changes, would consider urban noises "dangerous."

Knowing Camps and Pathways

Earlier I suggested that interpretation of forest images and acoustics requires experience (p. 150). Even then, one will never master everything. Interpretation implies knowledge, knowledge implies a sociocultural context, and society implies history. To reiterate: in the course of daily life one may accumulate many bits of isolated information, which will be meaningless without an intellectual scaffold to hang all of it together. That takes generations to build. Consequently, a lot of study is necessary. Batek attitudes to knowledge show a mix of self-effacing humility (admission of ignorance about one thing or another) and pride at self-mastery. There is a prestige complex involved in knowing things; people are quick to recognize adeptness and knowledgeability. A functional explanation of prestige is that it serves as a social impetus: though it may bring no material rewards, it motivates people to do what is necessary to acquire it. And so too with knowledge, it seems. In general, Batek seem genuinely eager to learn about new things and to improve self-knowledge.

When I first started fieldwork ten years ago, I found this eagerness most embarrassing. They watched closely everything that I did: how I walked or held a pen, what I wore, where I kept my tools and equipment, how the equipment works. (As this book's illustrations show, some have taken enthusiastically to the technology of photography.) Later I recognized that their reactions to me were just a symptom of their characteristic approach to knowledge. One boy was chastened when he discovered how much of another boy's botanical knowledge he did not share. Telling stories to me, people would exclaim at some particular individual's command of the repertoire. Discussing ritual knowledge, they would voice their desire to learn more. Young men would follow other men to learn hunting skills. Young women would attend a birth to learn midwifery from other women. No one was forcing anybody to do any of this. Yet, there is social approval too: *oh tahu? bənɛr* (s/he really knows) was a common expression of admiration.

Prestige may make knowledge desirable but, as in many egalitarian societies,[18] there is an absence of formal teaching structures. A further organizational problem stems from the high mobility of individuals and groups. If people are always moving about—dispersed in various locations around the forest, minus the technology that makes virtual communication possible—how is knowledge to be distributed? I've remarked that my own pleasure in *capturing* views betrays an impoverished attitude to knowing the forest: a desire for stasis; a misguided need for order and coherence. Misguided: because the forest is not known in an orderly manner by its people. Stasis is not possible for the Batek; similarly, knowledge tends to be distributed incidentally, often contextually. Much depends on place, on being in the right place at the right time.

What's important is to be around others. Individual volition and readiness to learn count for a lot too. People may schedule reunions in order to spend time with a chosen mentor, thus somewhat overcoming the "catch-as-catch-can" nature of knowledge acquisition. On an everyday basis, there is general recognition that in order to know the forest, one needs to be present in it, and to *cip bah-həp*, learning about the forest through observation and experience, imitating what others do, and developing one's repertoire of skills and sensibilities. Often, parents scold children who don't want to go to the forest and they will say that if they don't, they will not learn anything. ʔeyGk, explaining how he had learnt to hunt and dig, said that he did this by following each of his parents: "When my father went to hunt, I followed him. When I did not follow my father, I followed my mother." Only through long experience in the forest, can one become a dab hand at knowing it.

What then is in the forest that affords such experiences, that enables knowledge reproduction? This is the social-spatial context of the landscape. In chapter 3, we reviewed the significance of camps, pathways, and rivers. Their importance as cultural symbols invites us to consider their larger social significance as well. They are important places of history: not just physical settings or ideal images in the mind.

Life is a process of moving from camp to camp, all within the surround of the forest. Some campsites are well-used and have been around a long time; others are newly established. Thus one is always following on the efforts of the past and discovering new locations. More abstractly, camps are related to each other both in space and time. The distance between one camp and another may be no more than a few minutes. This might happen when two (or three) groups temporarily share the same foraging zone, which roughly corresponds to the valley of a river or tributary. More commonly in the itinerary of a single camp group, the walking distance between successive camps is at least an hour, if not two. The visitor traveling with them will come to hear of other campsites close by. Some of these will be used by other camp groups occupying a neighboring river valley. Often there will be stray visitors from these other camps. In the course of the daily round in the forest, the route might bypass or cut through old campsites.

The picture begins to emerge of a series of crosshatching and overlapping itineraries, of camp groups meeting other camp groups, of travelers meeting the traces of travelers before. The forest begins to look like any busy thoroughfare anywhere, full of paths, projects, schedules, and routines. Many things come together, intersect, and diverge in a camp, then—the physical tracks and trails and the biographies and memories of people (see chapter 5). These form social memory. As successive groups of people use the old camps, so does general knowledge about the geography and ecology become widely distributed. Given the heterogeneity of the forest ecosystem (a point we considered in relation to plant distribution in the previous chapter), talk alone—i.e., linguistically based knowledge—will not communicate all there is to know about a place. One needs to go there to discover and become intimate with it. Thus as long as people have camps to journey to and from, they can affirm their relations with the place and with those with whom they have shared these journeys, and communicate memories and histories, and teach.

What about the pathways? We have sketched out the framework of movement on trails, and their importance to land use and history (pp. 63–67). How do the Batek tell stories about their movements? For one thing, their stories would be

Plate 13. Young men visiting the camp at Tabuŋ, 1999. Their group had made a surprise appearance the night before.

laden with detail, detail that shows up in a fairly extensive and domain-specific lexicon: the topography over which one travels, the nature of the trail itself (say, whether it be clear or overgrown, how level or sloping, how straight or crooked), the fauna and flora encountered. Changes, both natural and anthropogenic, that have occurred since one's most recent experiences on the trail would be included. Strong emphasis would be given to the physical and intellectual challenges such as the ascents and descents, sidestepping around obstructions, climbs, river-crossings, search or pursuit, reckoning of location and direction. If any startling phenomenon occurs, or if one were temporarily disoriented, these would probably be highlighted, even exaggerated to a hilarious degree. Just as important, they would say why they had gone on that expedition and who else was in the group. Any puzzling observations would emerge. Such narratives are common in post-excursion memory swapping sessions. These sessions bring the information from a day's walk into the setting of the camp, thus generating more fodder for thought and future planning. Much later, they will recount a particular expedition in remembering moments of togetherness and shared adventures with certain friends or relatives.

Those *halbəw* are not just pathways to get from one place to another. They are imbued with local history. Moving on a *halbəw*, knowledge is applied to practical ends and shared with others and new knowledge in the sense of new discoveries about the forest will emerge. I suggested in chapter 6 that the ecological traces of the Batek's history can be discerned in the anthropogenic nature of the forest mosaic. Everyday actions shape the biophysical reality of the landscape: foraging and food-procurement practices affect the productivity, demography, distribution, and variability of useful resources; encamping opens up the canopy, lets in sunlight, and hastens the growth of different kinds of vines and trees; avoiding forest with old and tall trees, the people produce vegetational patterns that are amenable to their own dwelling practices. In turn, the continuity of the forest generates further, habitual action and sets the context for similar action in the future. Such actions are linked to events and experiences and experiences to individuals, as evidenced in the practice of naming places after people whose actions have left some imprint on others' minds. In such ways is a familiar environment produced.

The way the Batek recount their expeditions reveals much about the sorts of cognitive and social activities involved in walking down trails. Sometimes, like when they are headed for new resource zones and encampments, they are opening new pathways. As they walk, the Batek say, they actively assess and remember the characteristics of the trail and topography. Being able to remember the route is a source of pride and people become visibly perturbed and puzzled if they had lost their way in the forest. This new fund of knowledge will eventually, either directly or over the course of many secondary tellings and retellings, become part of what Widlok[19] calls "topographic gossip" which is then potentially available for everyone to use. For those who actually use the information to guide movement along a new trail, there is some computational process going on, as he or she complements the socially derived knowledge with personal observations of phenomenon observed or perceived during the journey.

The discovery of new camping areas, or arrivals in a hitherto unknown place, shows the importance of integrating the personal with the public. Often it is a leader working in consultation with the group who decides a route. The leader would usually have discovered the suitability of the new resource zone while engaged in routine foraging and collecting activities. As happened a few times during my fieldwork, other camp group members, trusting in the leader's skills and experience, will follow him if there are no conflicting and alternative objectives and directions. Once settled in a new place, I was told, people will hang back, will not take the initiative to launch work expeditions. They will seek out those who know the place and follow them: to fishing, hunting, collecting sites. Reciprocally, someone who is intimate with an area can often be spotted taking the lead: showing "newcomers" (*batɛk rɛ?* [lit. new people]) where to camp, where to look for resources, which routes to take, and the like. As na?Gk said, this was something she did as a matter of routine.

What, then, are the analytical implications? As should be clear, I am not dismissing the role of human agency and cognitive processes in knowing the forest, identifying it as a familiar place, investing in and deriving meanings from it, remembering its qualities and particularities, figuring out how to live in it. As Hutchins puts it, "human cognition is always situated in a complex sociocultural world and cannot be unaffected by it."[20] Moreover, the social organization of environmental knowledge is what gives it its historical and political contingency. What I suggest is the need to expand our definition of the "social" and to reexamine conventional notions of the environment as simply a space which people know through cultural lens—as something fundamentally external to society.[21] Landscape cannot be marginalized as just a brute material environment whose existence is independent of the activities of people in it. It is those activities, whether performative or cognitive, that give the landscape its meaning and in a political sense, its identity. To the Batek, the forest, marked as it is with so many salient and knowable features, is a place of history and community. So long as the landscape remains, and the Batek continue to return to those pathways and campsites and tell stories about their experiences, the possibility for developing their knowledge of the forest will persist.

To return to the nature of Batek knowledge, it is not something that resides only in the mind, to be handed perfectly to another mind in a continuous flow of language-like thought. Just consider the bodies of knowledge that I attempted to describe in this chapter—those derived from visual and auditory information. The fact that I can put all that into half a book chapter suggests that language-based communication is certainly part of the context of transmission. But I have hesitated a long time over the material and remain dissatisfied with it: I know I've only scratched the surface; I am frustrated by my inability to elicit more information from more people. Some dimensions of that knowledge, like the associations of indicator species with other species, are certainly label-based (what's often called declarative knowledge). But they were not revealed to me as a set text. Ladiy blurted out one detail in the course of talking about something else, ʔeyGk confirmed it and added another detail, taʔKadɔy reeled off a list, and on it went. I think this is how this kind of knowledge is learnt by the Batek themselves and that's how the process works in an oral society. Just as the Batek cannot predict from one camp to another who they will see there and by implication what they might learn there (and they resist making such predictions), so ethnographers can never expect to put the final word on their knowledge. Because there is always someone else with something new or different to contribute.

So much, speaking absolutely literally, is captured on the move, in process as the moving body journeys from one place to another, from one set of companions to another. Knowledge is not an invariant "thing" that is shut out from the ongoing biophysical processes; it is emergent in those processes. The spread of new knowledge from person to person, group to group, along with an endless process of analysis, debate, reorganization, and interpretation, ensures a level of innovation, creativity, and variation. Knowledge does not consist of a set of received ideas that is faithful to tradition but divorced from the concerns of ongoing life. Rather, it is constantly undergoing experimentation, change, and evolution. The basis for that change is, of course, the capacity of people to keep on returning to the environment, over and over again—to observe, analyze, remember, and discuss.

As I noted above, the Batek often lay the stress on participatory learning; removed, detached observation is frowned upon. After all, much of what needs to be known is *out there*: say, for example, the indicators of ecological change, the acoustic and visible signals, the tracks and traces of persons going before one. We should also note how fast details of the forest could change. If one habitually returns to a place, minute changes are recorded and one adjusts movements (and analysis) accordingly. But when one's pathways change, and one does not return for years to a spot, trails become overgrown, young saplings have become mature, trees have fallen and begun to rot, gaps are opened up, elephants and other animals have left new openings in the forest, bamboo stands have fallen over—the trail can be unrecognizable. The forest has a way of changing when you're not looking. Come back after a few years away and the path is no longer familiar. Abandoned completely, those places and pathways become *saŋyɛt* (empty of people) or *jɜm* (cold).

Look at Lus's route description in appendix B. It's typical of how the Batek give directions or narrate movements. They draw on precisely the sorts of ecological features that undergo rapid decay or growth: rotting trees, old campsites, fallen trunks, useful plants. So all such route markers have relatively short "shelf life." People roughly know the routes but they know that the landmark details can be quite fluid. They can tell you for example how many ways there are to get from Kɜciw to Tɜmiliŋ regions in Pahang, or from Pahang to Kelantan. But not many people would be willing to guide your passage unless they themselves have recently used the route. They need to keep going back; otherwise, they plain forget the details. A few times I asked people why they did not want to go to a particular place. They responded that they did not want to risk traveling alone because they had not been there for a while. They might get lost because they did not know what the trails were like. Much knowledge will consequently become "rusty" without continuous attempts to access and retrieve the finer details. This is one reason why the Batek are constantly monitoring not just where they have been, but where they have *not* been.

But change, of course, has a broader political dimension to it. Of this the Batek are perfectly aware too. As we've seen, they highly value their mobility and their need to live inside the forest. In part this is a reaction to politics: government agents regularly try to persuade them to give up their mobility and move to settlement schemes. But this the Batek cannot do, for to do so, would be to disconnect themselves not just from their place of sustenance and shelter, but from their history, their knowledge, and their knowledge of themselves.

Notes

1. Certeau 1984.
2. See, for example, Ismail 1995.
3. Diamond 1991, 84; Ichikawa 1998, 109, 112.
4. Puri 1997, 150; Whitmore 1997, 58.
5. Gell 1999, 239; Ichikawa 1998, 109.
6. The term is Kawada's (1996).
7. For some sonic, mythic, and ritual examples, see Kawada 1996; Levi-Strauss 1966. On a particularly virulent form of bird-hatred, see Song 2000.
8. See, for example, Anonymous 2002.
9. Eugene Hunn on Tzeltal (cited in Ichikawa 1998, 109); Ichikawa 1998, 108 on Mbuti; Puri 1997, 388 on Penan.

10. Ichikawa 1998, 112.

11. Ichikawa 1998, 111–12; see also Whitmore 1997, 60–64.

12. Roseman 1991, 35.

13. For a wonderful example from the Penan Benalui's hunting repertoire, see Puri 1997, 371–76.

14. Kirk Endicott 1979a, 65.

15. Lye 1994, 261–62.

16. See, for example, Dentan 1979, 23–24.

17. Kirk Endicott 1979a, especially pp. 70–79, 163–64; Lye 1994, 110–52, 273–76. On the dual personality of Gubar, there may be some continuities with hunter-gatherer religions worldwide: see, for example, Biesele 1993, 180–81 on the Ju//hoansi equivalent and Nelson 1983, who describes the Koyukon's Great Raven as "a magical clown, whose great spiritual power is mediated through an affable scoundrel" (p. 80).

18. Woodburn 1982.

19. Widlok 1997.

20. Hutchins 1995, xiii.

21. A point reiterated in countless social studies (for example: Flint and Morphy 2000; Hirsch and Hanlon 1995; Schama 1990) but still lacking full acceptance in the public imagination.

8

Changing Pathways

The morning of departure opened for me with a wake-up call from ta?Kadɔy. November 6, 1996: two days after Tebu's chat. The springboard for this as for so many other journeys was Was Yɔŋ. Specifically, the hearth of na?Ksɔ? and ?eyKsɔ?.

In those days, one of the men from the Semaq Beri village of Kg. Tiang, not far downstream, was paid a monthly wage by JHEOA to chauffeur his village children to school at Was Tahaɲ. He had charge of a long boat powered with an outboard engine, the standard mode of transportation on the Təmiliŋ. The day before I had asked him to stop by en route to the school; I needed to get to Was Tahaɲ, so as to board the 9 a.m. passenger boat down to the Wildlife Department jetty at Kuala Tembeling. From there Kuala Lumpur, the capital city on the west coast, was a taxi, or a taxi and a bus ride away.

All of this was well planned out in my head. However, I slept soundly that morning. Ta?Kadɔy hooted for me several times. When finally I stirred, the school boat was already waiting on the riverbank. I rapidly sized up the situation: me half inside a sleeping bag, not yet dressed or packed, boat down on the river, Semaq Beri children in uniforms expected at school very soon. I sent the kids to tell the Semaq Beri: *don't wait; I'll find another way to Was Tahaɲ.* After all, there was another passenger boat departing at 2 p.m. I could trek to the park headquarters well in time to catch it; I had made this punishing hour-long trek before.

The kids returned: *he said he'll wait for you.* Not for the first or last time I groaned ungratefully at the thoughtfulness of Orang Asli. *Well, if he's waiting then I've got to run.* In five minutes flat I had pulled on my pants, run a brush through my hair, rolled up the sleeping bag, stuffed everything into my backpack, heaved the bag onto my shoulders, jumped down from the house, and leapt into my sandals.

Throughout this whirlwind I could think to emit only one sound: *ʔɔ̃h*, the closest Batek equivalent to "goodbye." Until then I only dimly grasped the meaning of this word; I never understood why Batek professed such heartfelt sentiments when they *kãʔkɔ̃ʔ* (to say *ʔɔ̃h*). I was touched: many people were standing outside their houses or lean-tos. Some stayed inside; leave-takings are not easy for them.

A chorus of *ʔɔ̃h! ʔɔ̃h! ʔɔ̃h! ʔɔ̃h! ʔɔ̃h! ʔɔ̃h!* came from every direction. I ran through this gauntlet; turned my head left, right, left, right; *kãʔkɔ̃ʔ* back: *ɔ̃h! ɔ̃h! ɔ̃h! ʔɔ̃h!* Some were down at the riverbank. As I settled into the boat and the engine sputtered into life, I turned my head away from the Batek: I didn't want them to see my tears. Some distance away, I turned back and waved. I had understood the meaning of *ɔ̃h* at last.

Pathways: 1998–2003

Whose pathways will change? I know mine have. That logging road back in 1993; the tearful farewells of 1996; the later returns. To Batek, there's only one word to describe these journeys: *lew*. It means both to arrive and to depart. Like a path may be a path of ascent (*cəniwəh*) or of descent (*pənisar*) depending on which way you're going. At either end of the journey, you *lew*: every departure implies an arrival as every arrival implies a future departure.

What for me was once formless travel from one place in the forest to another has become a *route*: the pathways have led to this book. We can leave the Batek at this juncture or we can make another return. Ethnographic reportage always demands artificial closure. Yet, long after readers turn their attention to the next pressing story, the people and their pathways will still be there. Before winding up this journey, then, an updating is necessary.

I've always gone back to Taman Negara. Never again have I had the luxury of time or the chance to consult with Tebu on this book.

1998: a difficult year all around. There were dramas on the world stage: the year before the Asian financial crisis destabilized economies around the region. Commodity prices skyrocketed; Batek asked me why rice was now so expensive. I tried to explain import-export conditions. From village gossip they had also heard about the ousting, humiliation, and subsequent jailing of the Deputy Prime Minister Anwar Ibrahim. The Batek asked me how Anwar was. Some were worried that the political unrest in Kuala Lumpur would find its way to them. Well they were right; partisan flyers were posted all over Jerantut and there were political debates up and down the Təmiliŋ. But the Batek were not too concerned about the nation's leadership issues. Their thoughts as always are with military violence: they have always feared a reprise of the battles of the Emergency, which was largely a "jungle war." Would there be a war, they asked? Were there a war, I assured them, it would not happen here.

It was an incongruous setting for these discussions: all seemed so peaceful and business-as-usual back in Was Yɔŋ. Work on the access road from Jerantut (Plate 2; pp. 27–28) had stopped; budgets had run out. No newpapers or cyberspace here: no reportage of IMF policies, political chicanery, or street demonstrations distracted me. No sign either of the smoky fumes of forest fires that I had experienced in Borneo earlier that year, caused by a combination of El Niño-wrought droughts and plantation burnings. But of course plenty else was going on, many quiet dramas of everyday Batek life.

There was no question of harvesting seasonal fruits that year; they were focusing on rattan collection and tourism. A malarial epidemic had hit the Kəciw area: there were at least two Batek Tanum fatalities. ʔeyGk and his group moved away from Kəciw; several years would pass before anyone returned there. I continued to monitor everybody's activities. Tebu and his family was safe and well along the tributaries of the Təmaw. I took note of other groups and other locations; I detected nothing in the way of an upheaval.

Other personal issues consumed most of our talk. My own father had died; Batek had known of my worries over him. They had their bereavements too: naʔGk, once the center of a bustling sibling set and a close friend of mine, had died; so had the youth Taniŋ. And Blɛʔ—an unusual death from a snakebite. But the news were mixed as always. NaʔKsɔʔ was well though tired-looking; all the children were blossoming. A lot of other familiar friends were in the Yɔŋ area. We teased, laughed, gossiped, told stories, fell back into the old routines. I trekked into the forest to pay respect to Taniŋ's parents; I went up to Lal to visit Ladiy and Səraməh.

Next year another long list of births and deaths awaited me. Skinny, weak ʔeyKapey was dead, as was his little daughter. Both of Taniŋ's parents and their oldest daughter had also died inexplicably.

TaʔSudep and Tan, frail old men in their sixties and seventies, would surely have gone by now, I thought: but they survived another year. TaʔSudep even managed to father another child in the last year of his life; the baby did not survive. Tan was a different character: no gossip ever attached itself to him. He was long widowed, long since a great-grandfather; no more progeny would come from him.

ʔeyGk and naʔKapey, both widowed the year before, coupled up; another child was on its way. And how was Tebu, I asked? Still well, still alternating between Təmaw and Trɛŋin. His youngest, first known to me as a small boy with a catapult and a ready grin, revealed himself to be a growing youth with a first-rate knowledge of plant use. I rewarded him and a few others with a trip to Kuala Lumpur.

By 2003, the pattern had established itself. Five days opened up and I rushed to Taman Negara via a train ride from Kuala Lumpur and a taxi from Jerantut. Funds were pumping into the area; at last it was possible to take an ordinary taxi up the access road to Was Tahaɲ. For the first time since 1996, I experienced dissonance: I was losing landmarks. The taxi pulled into the village. And standing there were Batek doing what they've always done: picking up gossip and supplies from the Malay villagers.

Their camp was located at Tabuŋ on the eastern side of the Təmiliŋ. But it too had shifted location. The 1999 location, abandoned in 2000 and reoccupied in 2002, now contained burial sites. The Was Yɔŋ group, so large in 2001, seemed to have imploded. Following his wife's death the year before, the putative "headman" had retreated into himself. Even childcare seemed too hard for him now; one married daughter took over the daily sustenance of the youngest three. She had never borne children of her own and doted on her siblings.

For me, the most painful death was ?eyGk's. He had been my host, friend, and mentor in Kəciw. I had known him as a brilliant, versatile, and vigorous man; I did not expect him to die so young. I recalled all that we, together with na?Gk, had shared: the journeys, the forest trekking, yam digging, fishing, and hunting, the long nights talking about religion, singing songs, and storytelling. I found my thoughts drifting to their son Gk. I once saw him as a laughing boy tearing freely around the forest, the spitting image of his father, whom he clearly adored. With difficulty, I tried to assemble a picture of him now, as an orphaned teenager.

It all seemed so familiar. Yet I sensed that something had changed; but without the clarity that comes from a long-term stay, I could only guess at the implications. The Internet had finally arrived at Was Tahaɲ and entered Batek lingo. But the intricacies of telephone communications were still a little mystifying. One youth reported trying to place a call to me the year before: to say that the fruit harvest was a very good one and I should return for my share. He did not succeed. Over the years there had been visitors from Kelantan and Terengganu; reciprocally, more of the Pahang youths had made the long trek up to Kelantan. Folks from ?ato?, once unbudgeable from their settlement, were spending more time with their Yɔŋ relatives—and more time in the forest. They could take advantage of the tourism opportunities that had never come to ?ato?. The uppermost age strata that I had known was well and truly gone. In its place was an entirely new cohort of infants: and I knew exactly how old each child was. My place was also different now. Via my ties to na?Ksɔ?'s family, I had become a grandmother.

I too had changes of my own to report. After two years in Kuala Lumpur, with a detour to climate activism, I had relocated to a small town an hour away from there. The town is nestled at the foothills of the Main Range; the mountains surround us. From my front yard I look at this orographic skyline and think of what lies beyond: Kuala Lipis, from where I had first sought the Batek in 1993. I muse often about the cultural geography of these towns and their fading histories as colonial "forts." Batek, Batek Tanum, Semai—these were the Orang Asli who appeared infrequently in the shopping alleys of Kuala Lipis. How different here, with a Temuan settlement within town limits and villages just outside. Temuan are urbanized here and their belonging unquestioned. But they share space with others: Malays, Chinese, and Indians; long-term residents and migrants like myself. Not far up the road from where I live is the very place where the assassination of Henry Templer had launched the twelve-year Emergency; the nearby forests were for a long time classified as "black" (i.e., a security zone). Unlike the Batek, people in this town like a "clear" view: *keep the way clear so we can see if prowlers climb over your fence*, my backdoor neighbor urged me. The years of terror had obviously sedimented themselves in local landscape psychology, in ways that would repay further research, I thought. Recent state maneuvers have added new ways to think about landscape: within half an hour's drive from my home an enormous hydroelectric dam is filling up. Temuan communities were displaced; environmentalist concerns were rejected in the name of development.

On the Boundary

I hope this book lives up to Tebu's expectations. It is an ethnography of a landscape: the Batek's *hǝp*. The ethnographic contribution is to add to what we already know of it, the Batek's hunting-and-gathering way of knowing and using it, and the effects of degradation on forest peoples. Those effects are shared widely and not only among forest-dependent peoples; indeed, we find similar themes all over Malaysia as landscape fragmentation continues apace.[1] And as I've suggested we find them around the world too (pp. 19–23). I have shied away from a full-blown discussion of current theoretical arguments, whether of the nature of hunter-gatherer environmental relations, indigenous environmentalism and relations with the state, the categorization of the rainforests, or the very concept of landscape itself. Although those discussions have been helpful to place Batek conditions within wider frames of reference, more important to me is to let the ethnography speak for itself.

As I said in chapter 2, I make no apologies for the generalized nature of this ethnography. Of course I acknowledge that individual pathways may not always follow the normative ideal (chapters 3 and 4), that Batek are historical actors (chapter 5), that knowledges and perceptions are always variable and context-dependent (chapters 6 and 7). Most of all, that Batek landscape perception, like those of my new neighbors, are shaped to a large extent by global economic flows and governmental land use practices (chapters 1 and 3). What I've tried to portray throughout this book is the integration of old and new.

It seems important to try and address two faces of change and continuity. One: despite outside dismissal, at this point in time the Batek are maintaining this mix of the old and the new. I would describe their position as a "boundary status": poised on the fringes of yet not isolated from broader Malaysian society; neither assured of land tenure nor without land; neither fully assimilated nor isolated; neither in nor out of the shadow of the state. Two: some wider change *is* occurring and, at this historical moment, Batek are using their "boundary status" in order to carve out a vision of the future. Most important, then, is the political objective of recognizing that the Batek do have this vision.

I should consider their stories as *surat* or "letters," said ?eyTow in 1993. In the same way, I urge readers to consider this book as a letter: that communicates the Batek's environmental anxieties, that probes the practical and ideological reasons for those anxieties, that offers the Batek's aspirations of what *we* should do to "save the world." A letter that, however, has been channeled through *my reading* and therein lies my major anxiety with this book. I cannot claim that the Batek themselves would have framed their communication in precisely this way. Equally problematic—speaking authorially—is how to draw readers into this landscape, to make people want to get beyond their cultural filters and recognize the forest through the terms that now seem so self-evident to me. I have tried to show how the Batek engage with the forest and to highlight their perspectives. Whether I have succeeded is something that perhaps only the Batek can evaluate.

Substantively, I have responded to two major themes in Tebu's message, both of which suggest that the broader world has either forgotten the Batek or chosen to ignore them. The first is his reminder that "*heh*" (*they*, the Batek, and *us*, outside the forest) have a shared dependence on the forest. When the forests are liquidated in aid of external development, the forest people's share of resources declines. The parameters of that decline are what I have tried to suggest throughout this book. But, as Tebu reiterated throughout in his concern for the health, peace, safety, and indeed "soul" of the world at large (see, for example, pp. 27, 34–36, 38–39), we also share the effects of degradation. Tebu wants us not to forget our shared predicament—they haven't forgotten, not their linkages to us, nor the reasons why the forest has become so vulnerable and unstable.

So, and this is the second guiding theme of this book, one way to keep remembering is to learn from them. Among many possible lessons, those I have emphasized are inspired by Tebu's final passages (p. 39), when he underscores the need to live in a certain way in order to hold on to the memories, the knowledge, and the environmental consciousness—what, earlier, I described as an "ethic of care." They have a strategic place in the forest. They can use it aggressively or conservatively: the choice does exist. But the aggressive way, as Tebu reminds us, will only "kill the world." The problem, of course, is that their conservation choices aren't really going to mean that much in the larger picture. As Tebu recognized, it is those who "live tame" who are most responsible for degradation. Hence the need to send out the message to those who have wealth, power, and influence. For if the current trajectory does not change, their way of life will become ever more unhealthy, precarious, and difficult to maintain. With deforestation, forest cover is transformed to "tree cover" (pp. 8–9, 22, 29–30); waters that should flow freely along their natural courses are inhibited by siltation (pp. 31–32, 36–38) or channeled to hydroelectric dams (pp. 10, 22, 38); the seasonal rhythms of foraging and collecting are disrupted by habitat change and species depletion (pp. 112, 115–16, 144); familiar landscapes are mowed under and brought within the ambit of townscapes (pp. 7–8, 97–99). What from one perspective are environmental problems assume political importance, for the Batek face a critical challenge: to remain in the forest, despite its inherent troubles, or to leave the forest with all its ideological and social implications.

One implication is suggested by Tebu's statement that Batek will "show the way." By showing us how they live and the principles that give meaning to life, they offer an alternative vision, which is contrasted to that promoted under the current development regime. It is a way of knowing and dealing with the forest, which has its own epistemological integrity and logic. I don't think he suggests that this way of life is appropriate for everyone but that it is an example of what *can* be done. The main pedagogical value for us, as stressed in his message, is that it promotes "love" of the environment, an appreciation of what gives it its "soul," a heightened consciousness of limits and awareness of the impacts of degradation. If they were to select the government-approved route to "modernity," *we* also feel the effects because the people who can "show the way" will no longer *be*.

The major guiding motif of this book is the very existence of Tebu's message. The message ends with an argument for "cultural preservation," but it begins with the need to communicate. And it reflects, I suggest (pp. 117–19), the rise of a claim to environmental stewardship. Such a claim does not arise from nowhere: its content is rooted in traditional cosmology; its context is the degradation of the forest and the world. And it is a response to what the Batek see as the connection between forest loss and public ignorance and denial of their voices and knowledge. They have arrived at a juncture where, independent of outside intervention, they are shaping an environmentalist perspective on degradation (p. 23). This says something about the state of the world today. No longer is it possible for communities like the Batek to take a laid-back attitude and wait for governments to deal with problems. They see fit to mobilize themselves because it seems to them that their governments are not seeing the problem (pp. 20–22). We need, then, to understand the Batek's position against the background of their practical activities and forest representations *and* their political economic relations with external forces. Cosmology provides the intellectual scaffold—albeit a flexible and adaptable one—for the Batek's representations while their practices are integrated as a hunting-and-gathering production system; there is also a mutually reciprocal relationship between ideas about actions, those actions, and the material conditions of the forest. Furthermore, the forests (and other landscapes) are not just biological phenomena; local use practices and the larger political economy interact to produce a landscape mosaic. Different use strategies produce different kinds of microlandscapes. Some environmental changes are acceptable; not all are destructive.

With the foregoing as the underlying themes, this book has been a straightforward ethnography: an ethnography of a landscape whose meaning is derived from the activities of the people who dwell in it. So, rather than document the pace of decline, I have focused on what persists among the Batek and what they'd like us to know about their way of life: their sense of the forest as home, their way of making home and marking belonging, the landscape of the forest, how it is classified in myth, belief, and ritual. Most importantly, I focused on how the Batek identify themselves: as people of the forest, and people whose identity is constructed a certain way by outsiders. For a major grievance is that we have not yet recognized them, their knowledge, their rights, and their claims. Rather, outside perceptions are largely negative and, wherever benign, patronizing (pp. 23–24, 96–97). Ultimately, I do not believe it is possible to demolish such perceptions on epistemological grounds alone: they are too embedded in culture and history and legitimized by too many political and economic justifications. Yet it would be absurd not to try: that is my anthropological objective.

Changes in the Forest

Implied in Tebu's critique is that resource appropriation is based on narrowly commercial perceptions of the forest, which we may contrast to the idioms and images that he uses (see, for example, pp. 30–31, 36–38). These suggest, to reiterate, that the forest is not just a source of food and a repository of biological richness, but has a life, a soul, and makes humanity possible. In contrast, for many outside the forest it is just empty space standing in the way of development, devoid of social meaning. Whatever meaning that exists is derived from external paradigms, whether in terms of ecosystem preservation, species and habitat conservation, commercial forestry, or governmental land-use plans. These contentious paradigms are betrayed by the labels attached to any block of forest; i.e., whether it be called a "national park," "state park," "amenity forest," "wildlife sanctuary," "forest estate," "timber reserve," "production forest," "statelands," and so on. The largest contiguous tract of forest in Batek territory goes by the name of Taman Negara: "*national* park." That name does not recognize the Batek's existence, let alone their histories, stories, language, or land claims.

It may be intuitively obvious that the Batek are going to define the forest differently. Yet, as I pointed out in chapter 3, they do not put in so many words what *forest* is. In probing the linguistic references, I found, rather, that *forest* is a multivalent category, depending on context and time (see, for example, pp. 50–54, 67–73). One important context is social: the meaning of *forest* depends on who is living in it. And for the Batek, perhaps the most important meaning in the present context is that the forest is important to their identity: remove them from the forest and they wouldn't be Batek. That's what they've always said to me. Reciprocally, the forest also needs to be understood in relation to its people. In contrast, state discourse tends to represent problems with either/or dualisms: environment *or* development—development concerns must take precedence over environmental protection. So too some brands of environmentalism: environmental concerns must be more important than people-concerns. The Batek position seems different: people must be put in the same frame as environment. Kill the forest, and the impact will be felt by people.

Which all seems in line with their general ambivalence toward the forest (see pp. 50–51, 89). It is by turns positive and negative, safe and threatening. Were we purists we could argue that this ambivalence is solely the product of today's conditions. As I argued in earlier chapters, that is probably not true. The seeds of anxiety are found with a close reading of the cosmology. This suggests that the strength of the cosmology is not only to give meaning to behavior and environment but to *anticipate*. That is, it is flexible enough in structure that it can be adapted to change with changing conditions. One implication is that Batek are not pinioned to the past; their perspective is not an artifact of tradition, of primitives panicking over change. They are not responding as "primitives" who are just "not exposed" enough to the benefits of development and are too "shy" to join the world outside (the terms within quotation marks are common staples of statecentric discourse on Orang Asli). Contrary to external perceptions, they do not feel that it's contradictory to be *mɔden* (modern) peoples living in the forest. I too went into the forest with my preconceived notions. Somehow or other (in 1993) I thought it was significant that the Batek wear wristwatches, have radios and backpacks, put on bras and Tshirts when visitors come, and so on. These are only surface symbols that can and do coexist with a life based on mobility and hunting and gathering.

In the traditionally linear view of history and modernization, change would always be bad; people would forget their culture, they would become assimilated, they would have all kinds of new needs and wants, and they would lose their religion and moralities. I won't claim that this process is not happening at all among Batek. I can only assert that it is not what I have found among Tebu and his protégés and colleagues in Pahang. What this shows is that the pathways of change cannot be predicted. People respond to new opportunities for different reasons and in different ways. The way followed by the Batek of Pahang is to make a clearer stand of their ties to the forest and their grievances that these ties, and the sustenance, knowledge, and experience they afford, are not given their due credence in policy formulation.

The whole thrust of Tebu's communication is to send the message to those who do have legal powers to reconsider what *they're* doing to the forest. Other than the physical evidence of degradation, which everybody can see and experience, there are also links to the government and economic relations with forest margins entrepreneurs and neighbors, all of which ensure that the Batek are kept well-informed about the challenges facing them. Tebu's is not a cry in the dark; he has a shrewd idea of what's going on and what alternate routes to the future may promise. And rather than resist change, he *wants* change. He is asking for a less exclusionary approach to government administration. Like every other Batek who has discussed these matters with me, he wants recognition that alternative modes of living and acting are valid and valuable and important to the future. Batek reject most of what they hear from government officials: for example, that they are "living wild," should settle down, put down some roots, plant permanent field crops, be easily reached by officials, or adopt Islam. Of course just as Batek perceptions may be variable from person to person, so too is the government not a homogenous entity. Not every government official espouses this medley of ideas to the same degree or kind. Notably, Islamic conversion is promoted only by designated officials. But such themes are reiterated over and over again, and from these the Batek have distilled an official government attitude, which they feel strongly about.

Now, in the best of all possible worlds—lands and resources intact; no external claimants; government promotion of socio-cultural diversity; adequate security of tenure—oppositional discourses would not arise. But in the last few decades, the Batek see themselves losing forestland and, relatedly, cultural autonomy. They have grown more alarmed. They know the forests are vulnerable to predation (see chapter 5) and now they can see exactly how frail and diminished the forests are becoming. At the same time, more external sectors have staked their claims on the forest: logging concessionaires, plantation owners, dam-builders, tourism outfits, private developers, government agencies. These are powerful forces. The Batek don't have the population numbers to counter these people. They know what these claims mean for them: loss of forest.

Struggling for a Place

In Batek landscape, only Taman Negara has statutory protection. It has a special significance: one of the final remaining stretches of unbroken lowland forests in the Peninsula and a haven for biodiversity protection. The Batek are publicly known as the park's "aboriginal population" but this is as far as the recognition goes. It is biodiversity and not people that the park is associated with. That is, though the park *as the Batek's home* has a local character, its wider reputation rests on its national and global status. There is little symbolic indication of the Batek's presence in Taman Negara. Topographic maps clearly mark out the main river systems but the tributary systems between them might as well be terra incognita. Never documented in official releases is the territorial issue: that most of this national park sits astride Batek territory. Or, as Batek said, "We came first."

Most of the Batek landscape, as we have seen, is either clearcut or logged-over, transformed or fragmented. Fragmentation, as we saw in chapter 5, does not necessarily make a forest patch out of bounds to the Batek; it can still regenerate (ecologically) and it still has integrity and value (culturally). Politically, it is still part of the territory. One can deal with this landscape reality in two ways: to focus on the inside or the outside. As I pointed out in chapter 7, mine is the liminal position, neither in nor out of the forest (pp. 148). Largely I have chosen to go *inside* the forest rather than outwards toward, say, the spheres of environmental decision making in the metropolitan centers. This may have communicated a different sense of place than what we normally associate with the desolation of deforestation. Indeed this is the aim—to show that despite it all, the Batek continue to find the forest a meaningful place to live in and that so long as there are vast stretches of forest like Taman Negara left in the territory they can derive sustenance from it. The political reality is that they are losing ground and we must deal with the implications of that.

In thinking of the forest, then, keep in mind that it is also a site of struggle. Politically, that struggle is over land and resources. Forests are no longer entirely local landscapes; they are resource landscapes (as, before, they were also military frontiers), and anything that is not claimed privately or titled to groups belongs to the state or federal governments. Local-use values tend to be pushed aside in this trajectory, with the state often accused of not shouldering its responsibilities to manage resources equitably. As suggested in the introduction (pp. 6–7), there is a close relationship between forest exploitation and political patronage in countries like Malaysia. Resource exploitation at its best converts natural capital into social programs for the greater good. At its worst, it gives another avenue for economic rents to be appropriated by a politically influential minority. It is questionable to what extent local peoples benefit from these rents. The percentage of Orang Asli living below the official poverty line is highly disproportionate to their numbers.

Monetary standards allow us to evaluate dispassionately who benefits and who loses from forest degradation. But they do not capture the full extent of forest peoples' losses. Part of the objective of this book is to draw in broad strokes different kinds of loss: knowledge declarative and perceptual; resources upon which livelihoods depend; history and cultural memory; the meanings of landscape; the basis for identity and social reproduction. When we consider these and everything else that people derive from the environment—good health and sanitation from having adequate access to nutritious, well-balanced meals and clean water, psychological well-being from having a sense of belonging, knowing that life is going well, and having the support of a community that is not falling apart—then it becomes impossible to reduce values to monetary and other forms of quantitative standards. These quality of life considerations provide a more meaningful (if hard to count) measure of how the liquidation of forests makes people poor.

But though government administrators might sympathize with this, and do feel the need to gazette land (for whatever reasons), they are quite reluctant to give managerial and ownership status to those who belong in and to the forests, even through benefit-sharing or co-management agreements. One common feeling among administrators is that local peoples are naïve and cannot be trusted *not* to sell forestlands they have title to. Although this is valid enough, in view of the growing cooptation of local peoples by outside commercial interests, we do not find similar caution against granting timber concessions to outsiders who feel no personal responsibility toward the forests or even the valuable timber in them. Their orientation is not toward the forest itself but towards the larger conditions and opportunities of the global commodity markets. Nor does granting indigenous title to land rule out the possibility of government agencies retaining a superordinate stewardship role over the future of the forests in question. There is a wealth of co-management regimes that could be adopted.

In governmental presentations, development is usually presented as a "good" and its benefits wholly self-evident. From this perspective, those Orang Asli that are under state administration—like the sedentary Batek of Kelantan and Terengganu whose settlements receive much government attention—are moving in the approved direction. They are on the way to giving up their "traditional" ways. For ethnic minorities, however, development has eroded their resource base and led to more state paternalism and control; the general result is the loss of local autonomy and subsistence security. As we saw in the discussion of Semaq Beri dependence on wild yams (p. 128), even when settlements are provisioned by government agencies, subsistence security is never guaranteed. Yet a stock response of state agents to local critiques of development-sponsored changes is that the people are too primitive to know what is good for them.

Taking the most balanced view of the matter, I believe that many government officials I have met are genuinely well-meaning; they cannot see what is the value of the traditional life (as well as being unself-conscious over the distinctions between tradition and modernity) and therefore feel it as a duty of the state to actively intervene and improve local peoples' lives by changing them. Moreover, officials are not unaffected by criticism. There *are* officials who feel genuinely frustrated because their agencies and departments cannot do more for the people, when programs fail to deliver the promised benefits of development: land security, educational opportunities, higher standards of health, sanitary living conditions, arable land, job opportunities, and so on. Such failures lead to cynicism and distrust among the "clients" (this word is used in the JHEOA's official website statement of objectives, its "clients' charter"). Nor would I criticize those Orang Asli (and other indigenes) who have risen above the conditions of poverty and discrimination to achieve high educational and professional levels. But their examples cannot necessarily be held up as a model for all indigenes to follow. Rarely valued in governmental discourse is the question of choice. Not all Orang Asli choose to follow the dominant pattern of development. There is space, I mean, for maintaining alternative conceptions of a meaningful life and the Batek's is one such conception.

The governmental presentation assumes that there are only two groups of people in the world, those who appreciate development and those who don't, and those who don't need to experience it more in order to join those who do. But the Batek have already seen the world from their perspective. They are every day confronted by countless images of landscape degradation. They are already exposed to the benefits of development. They have little choice but to advance a position on what this degradation means, drawn from their knowledge and sentiments of the place and its history. It is this knowledge, its cognitive and imaginative dimensions, that has formed the substance of this book.

Constructing the Landscape

This book has shown that, just as people's biographies are grounded in location and geography, so is landscape biography—its history—intimately connected to the people's capacity to keep in touch with different places. With growing degradation, there is the possibility that people will become more estranged from their geography of knowledge and that, ultimately, landscape lore becomes just lore, history without a place. A major theme was that the Batek's environmental representations are both a concrete expression of local knowledge of place and a response to the sweeping changes that are altering the landscape. The value system, drawn as it is from an ancient cosmology, has expanded in scope. So has the scope of the audience grown. Tebu's message is a tactic of reversal: from his perspective it is not the forest people who need more education and exposure; it is us, the targets of his message, who are endowed with ignorance and who need to hear the stories of the forest.

The Batek's "silence" is part of the long history of official dismissal. In official construction of the environment, local communities present a "problem," either for governance or against forest degradation. Their presence in the forest is deemed to be a threat to sustainability and a problem that prevents them from "progressing." The "problem" may be constructed for short-term ends (to facilitate land and resource exploitation) or because officials genuinely believe that living in the forest offers the people no reasonable future. For the official position to make sense, then, the forest cannot be perceived as a dwelling place, it has to be peripheralized and conceptualized as culture-free. The problem is that, if one thinks of the forest as culture-free, then one is forced to think of the forest people in the same terms:

wild, stupid, ignorant, primitive, and brutish. Further, this position seems to be scaffolded by the Malay distrust and abhorrence of the mobile way of life (chapter 5). It becomes natural not to engage in the forest people's conceptions and somewhat ridiculous to do otherwise. The lack of attention to the Batek's voices and concerns, then, is hardly surprising.

Tebu's message challenges that dismissal. The Batek recognize that it is not quite possible to address government representatives directly. They want to sit down and discuss the future of the world with those who can arrest degradation— they want to *pakat*, to use Tebu's word choice—but they cannot just step up to a microphone and say that. In fact, there is no microphone proffered so they have to look for other ways of voicing their concerns. If the offer to *pakat* were accepted, it would be an important political development and recognition. But we need to be careful: the terms and premises of any such discussion will inevitably be loaded against the subordinate group. They would have to justify themselves using the terms and premises of the outside world rather than the other way around: as a group of Orang Asli leaders did, when they declared that they are "not anti-development."[2] By using that term they have already accepted the official discourse that Orang Asli should aspire toward the state-sponsored model of development. Would there be room in such settings for Tebu's vision? Recall how he admitted, "There are Batek who want to be rich." Because, he said, "It's not easy to live like this, we suffer." Yet to live one way rather than another is a conscious choice, which those like him have made: "But this is preferable to killing the world, like life outside the forest." Is there a room in the state for such a discourse, for public admission that despite its inherent disadvantages, there are important values in forest-dwelling ways of life?

When intellectuals like Tebu reflect on the implications of degradation, they do not argue in the expected way. Foreseeing a world in which they might change or life in the forest become more fragile and difficult to sustain, we might expect them to demand protection of land rights or for more effective provision of government services. While on the one hand, to recall comments made in chapter 1 (p. 17), they acknowledge that they are "poor" relative to folks like me, they do not take this to its logical conclusion, that they should *not* be poor while others are getting rich or that they are *entitled* to an equal share in development benefits. Self-perception of "poverty" is, in other words, a descriptive statement rather than grounds for a moral discourse. I have, for example, no record of Batek using the Malay word for "rights," *hak*; the language of rights and entitlements has not entered their vocabulary. Arguably perhaps its absence suggests political naïveté. On the other hand, it is also plausible that they have not accepted or assimilated the language of development. Because, were they to demand equal rights or entitlements, they would be acknowledging that they *should* be like those people outside the forest who are, in Tebu's words, "killing the world." The challenge, then, is to understand what forms their sentiments take and how these are expressed.

Let's return to the boundary. As expressed by Tebu, the boundary can be a source of strength and cultural vitality: so long, that is, as the forest remains. If there is continuity of place, there can also be continuity of experience and knowledge; i.e., there is a physical connection between worlds present, past, and future. People who don't live there don't have this knowledge—they have no memory to move forward with—and so from the perspective inside the forest outsiders are the feeble and powerless ones. The forest, as I've been told innumerable times, has its own stories, its own norms, its own problematics and the people of the forest the ones who best know it and how to cope with it.

To be fair to the Batek's critics, this sentiment is widely shared. Malay villagers and even government officials, when they exclaimed at my "bravery" (Malay *berani*) in doing fieldwork in the forest, often assured me in the same breath that "you'll be safe with them [the Batek]." However, while the Batek's superior knowledge is easily accepted when the topic of concern is forest survival, the idea that that knowledge might have something to contribute to the urban-industrial world's way of doing things, specifically of treating and allocating natural resources, is controversial. Politically, Batek environmental knowledge has never been explored or treated seriously by planners. Just as in talking about life, health, and death, Tebu is appealing to universal processes of growth and decay, in Batek view the problems that beset their environment beset ours. Tebu says they want to talk: that means they want their knowledge recognized and they want to participate. In essence, if we continue to relegate their knowledge to the periphery, we will die along with them.

Conceptually, my study has demonstrated two issues. First, the customary dichotomy between culture and nature, society and environment, does not take us far into the study of local knowledge. Analyses that begin by taking people *out* of the environment, and thinking of that environment as a human-free zone, are also representations that deny both landscape biographies and the biographies of the people who live in them. The Batek see themselves as integral members of their environment and do not have language even to describe "wilderness" qualities. Second, that local knowledge is consistent with change. The Batek are always looking out, comparing, contrasting, grouping and classifying ideas, people, and phenomena, thinking about what it all means in relation to themselves. As the landscape changes, there is just a hint that their knowledge base also begins to change. As trees are cut down and forest opened up, the visual qualities of the environment change. New things come into experience, and this would be reflected in local history as well. New actors come into the scene, giving the Batek a more developed sense of the global implications of their ideas. Thus the reality is that Batek lives do not fit the popular conception that they are "primitives" lost in time. It is worth challenging these outside constructions, if only to craft a more imaginative and sensitive understanding of the effects of forest degradation.

Notes

1. See, for example, Kua 2001.
2. Zawawi 1996a.

Maps

Map 1. The states of Malaysia

Map 2. Distribution of Orang Asli groups. Adapted from an original map by Geoffrey Benjamin.

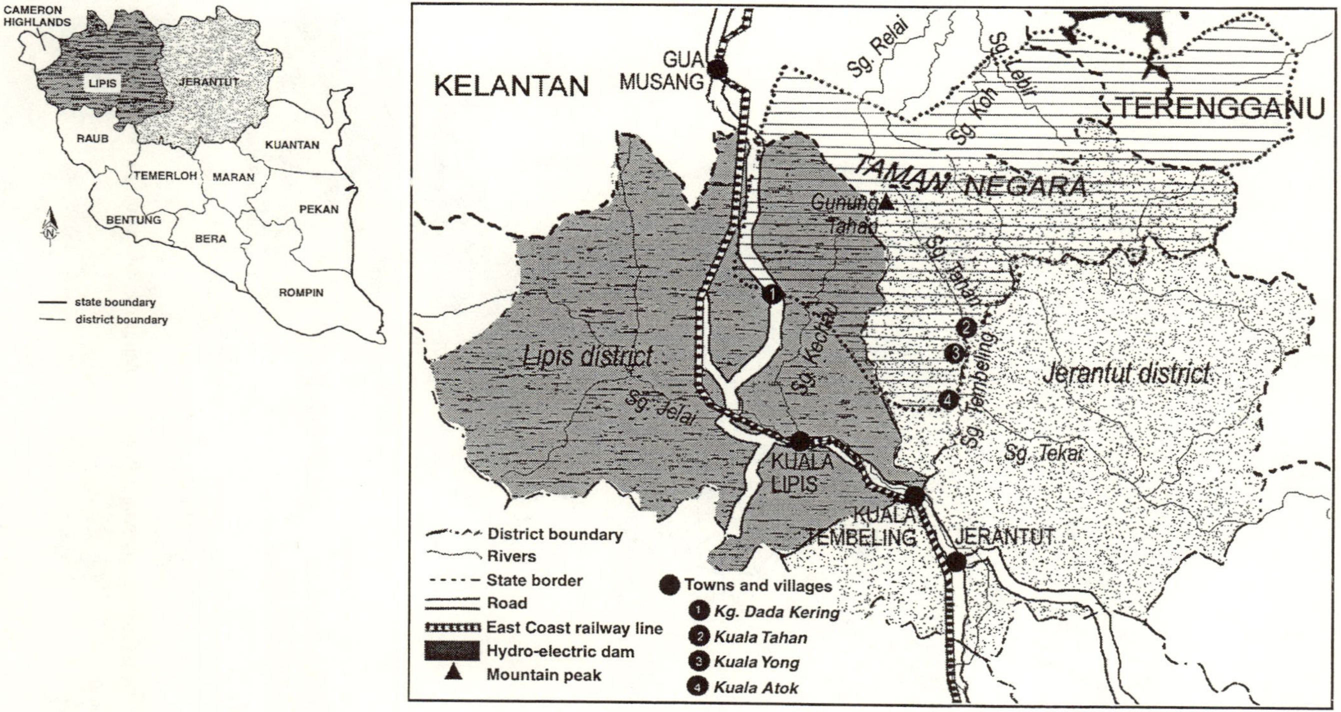
CAMERON HIGHLANDS
LIPIS
JERANTUT
RAUB
KUANTAN
TEMERLOH
MARAN
PEKAN
BENTUNG
BERA
ROMPIN
N
state boundary
district boundary
KELANTAN
GUA MUSANG
Sg. Relai
Sg. Lebir
Sg. Koh
TERENGGANU
TAMAN NEGARA
Gunung Tahan
Sg. Tahan
Sg. Kechau
Lipis district
Sg. Jelai
Sg. Tembeling
Jerantut district
Sg. Tekai
KUALA LIPIS
KUALA TEMBELING
JERANTUT
District boundary
Rivers
State border
Road
East Coast railway line
Hydro-electric dam
Mountain peak
Towns and villages
1 Kg. Dada Kering
2 Kuala Tahan
3 Kuala Yong
4 Kuala Atok

Map 3. Lipis and Jerantut districts, Pahang

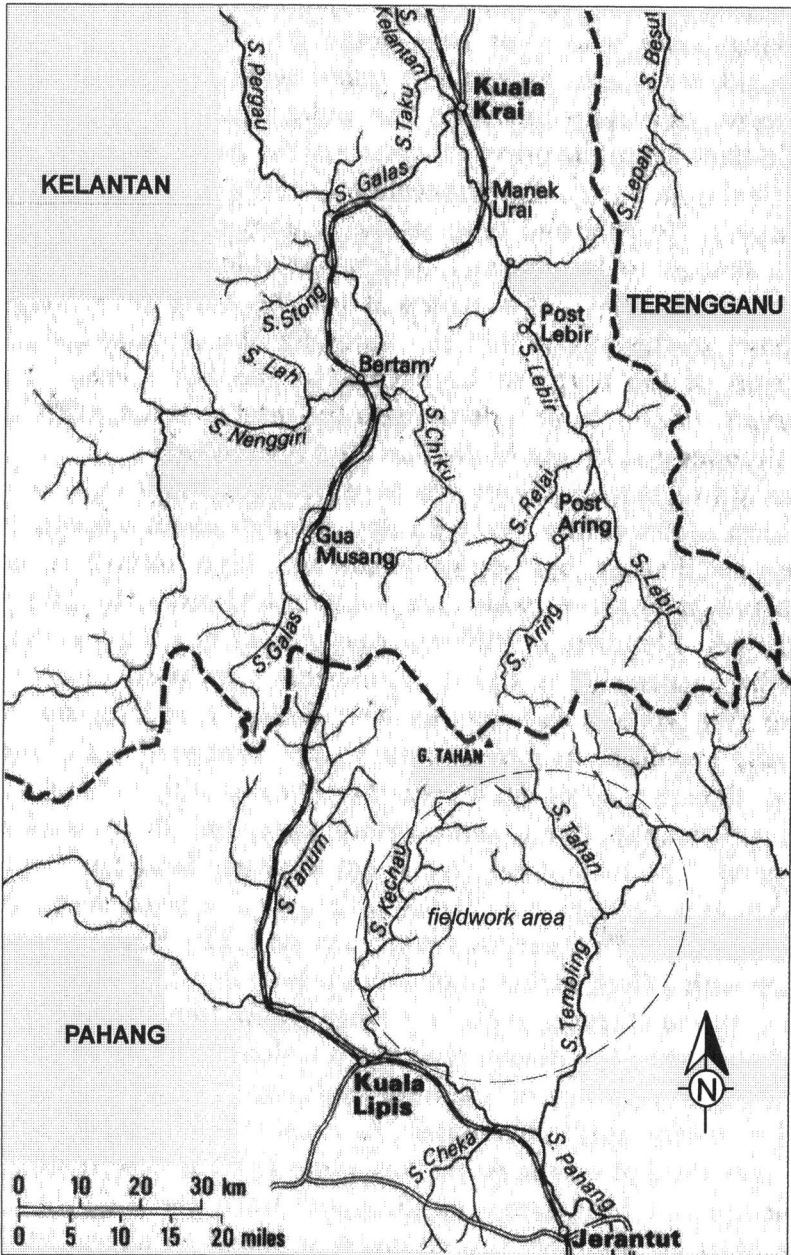

Map 4. The Batek territory in Pahang, Kelantan, and Terengganu. Source: Kirk Endicott.

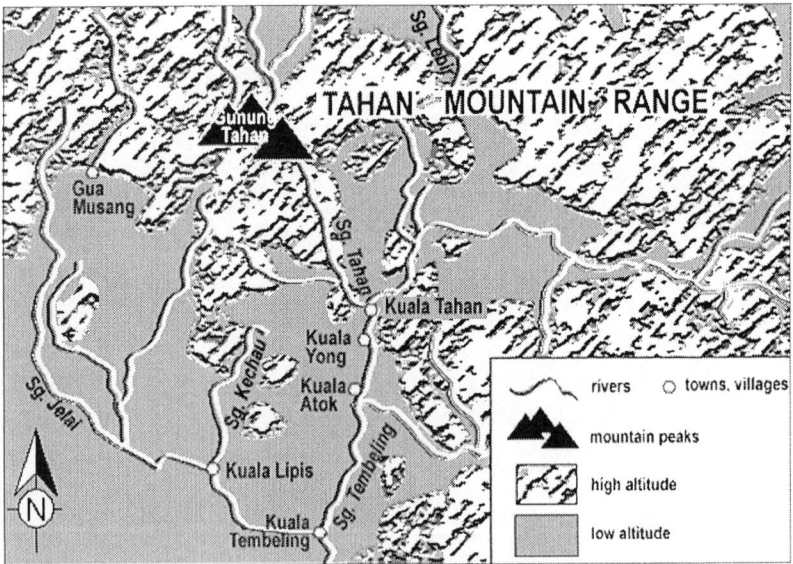

Map 5. The fieldwork area in relation to the Tahan Mountain Range. Map adapted from http://jpsscada.moa.my/phgmap.htm. Issued by the Department of Irrigation (original scale or digital resolution unknown).

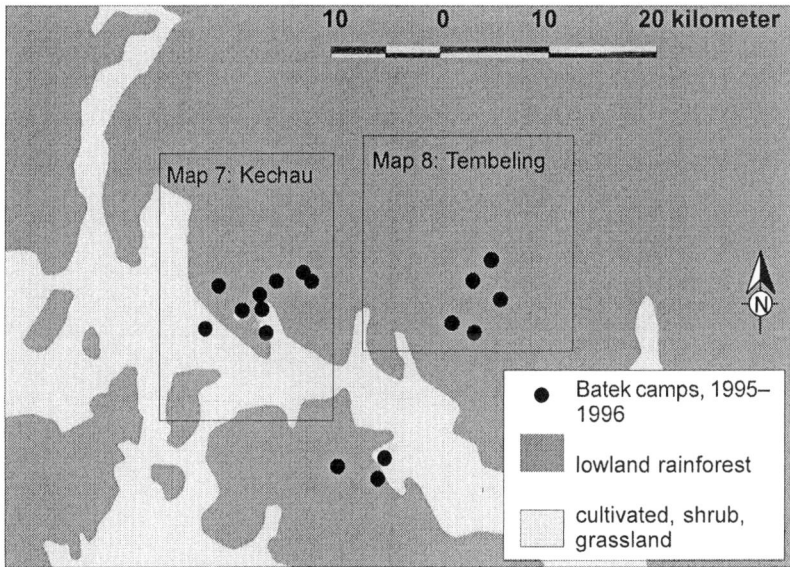

Map 6. Batek camps in Pahang. Vegetation map sourced from the World Conservation Monitoring Service (1993). Original scale 1:1000000.

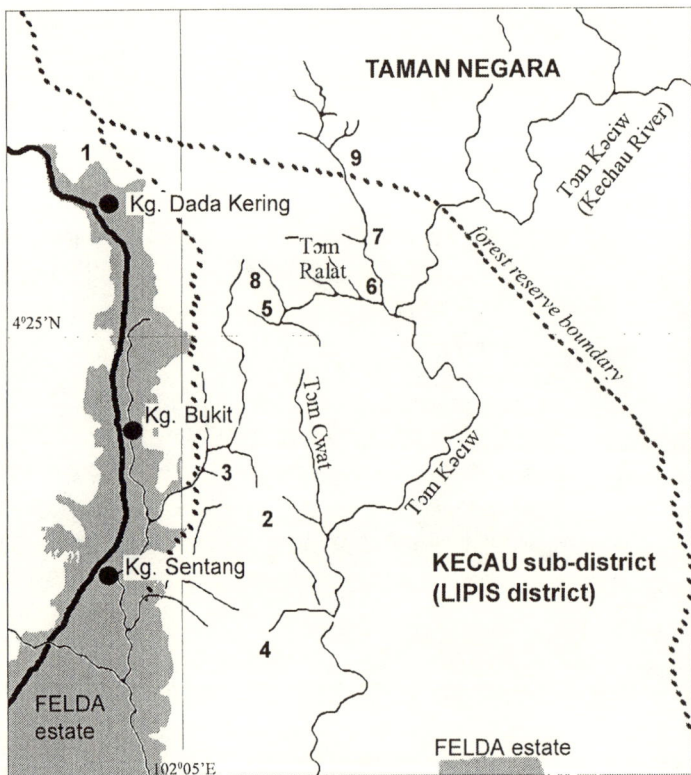

TAMAN NEGARA

Tɔm Kɔciw (Kechau River)

1

9

Kg. Dada Kering

forest reserve boundary

Tɔm Ralat 7

8 6

5

4°25'N

Tɔm Cwat

Kg. Bukit

Tɔm Kɔciw

3

2

Kg. Sentang

KECAU sub-district (LIPIS district)

4

FELDA estate

102°05'E

FELDA estate

▨ under cultivation ▬▬ surfaced road ● Malay villages

List of camps

1	Marɛm (Batek Tanum village)	6	Was Cənrɛt
2	Kəciw	7	"tom naʔ Kəciw" (name not recorded)
3	Taŋuy and Tɔləm	8	Tɔm Bəsut
4	Buməkəl	9	name not recorded
5	Rinərəm		

Map 7. Locations of Batek camps in Kechau, recorded with a GPS receiver, 1995–1996 (not comprehensive). Base map source: the Malaysian Mapping Department's 1:50000 topographical map.

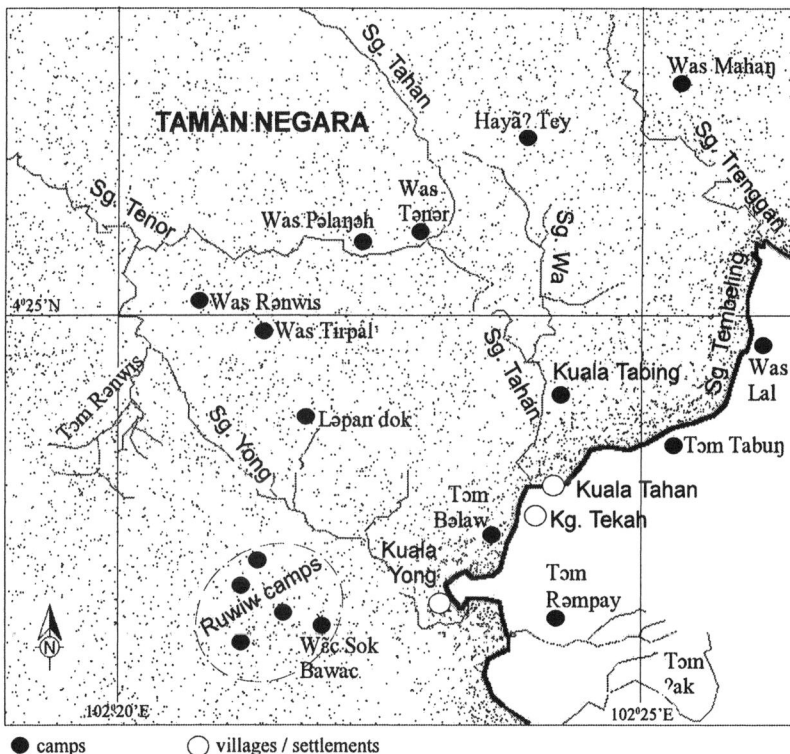

● camps ○ villages / settlements

Map 8. Locations of Batek camps in Tembeling. Digitized from 1:50000 topographical maps issued by the Malaysian Mapping Department. Campsite locations determined with a Geographical Positioning System receiver except where dense forest cover obstructed reception.

Appendix A

Tebu's Message: Transcription and Translation

The first part of this appendix reproduces the transcript of the relevant portions of Tebu's message (digressive sections are excluded). The second part contains my word-for-word translation. The third is the "free translation" that was presented in chapter 2.

As noted, I did not use a tape recorder. I scribbled these notes in a small writing pad as Tebu was talking. The transcript shows the effects of this: a combination of stenography with summaries and interpolations in English. In some cases (for example, the last line of paragraph 3), the sentence is not grammatical; it is just a collection of "keywords." The morning after the interview, I checked the notes for flow and legibility and added brief notes on Tebu's intended meanings and bodily gestures, as well as to fill in gaps; where these appear in the transcript, they are in English.

All the English insertions are in italics. Where the English words are part of a sentence, they are bracketed off from the Batek portion. In Part II, the English portions are not reproduced; they are indicated by place-holder brackets, as in [].

I have noted the significance of using the pronouns *heh* (we [inclusive]: we-all *and* you-all) and *?ipah* (we [exclusive]: we-all *but not* you-all). These are abbreviated below as "we [EX]" for *?ipah* and "we [IN]" for *heh*.

Each paragraph is numbered; the same numbering is followed in all three sections.

Part I: The Transcript

1. *[Take away the forest]* dəɲaʔ habis. *[Want people to know]* dəɲaʔ habis. Sudah neŋ kayuʔ.

2. Bilaʔ gən bom, Gubar manah ʔujan. *[Gave the example of the road construction on other side of the Tembeling. Said the reason why there's been so much thunder and rain lately is the dynamiting of the hills.]* ʔipah ʔiɲit.

3. *[Don't take away more forest]*—diʔ hares oɲ. Heh cariʔ makan samaʔ *[Want to pakat; in other words, we all use the forest, let's sit down together and talk about how we can all cariʔ makan].* Neŋ boleh radiʔ səpatut yaŋ heh cariʔ makan.

4. Tɔm Tǝmoh neŋ kayuʔ. Kǝlapaʔ blaʔ-blaʔ.

5. ɲawaʔ ʔoʔ gɔs ʔatɛs kayuʔ. *[Compare forest to* ʔurɛt.]

6. *[Because this land is]* polaw *[how can you]* tahan teʔ.

7. Kǝ-sɛn heh ʔaman neŋ rugiʔ dǝɲaʔ.

8. Batek halaʔ ʔoʔ kǝdǝh: "kǝlaŋes teʔ ʔoʔ diʔ."

9. *[Superhuman beings]* gǝn ʔiŋit gǝn haʔip.

10. ʔoʔ kǝsian hakaʔ.

11. Gǝn sayɛŋ *[so they tell us about things]*.

12. ʔipah kǝjeŋ—ʔay lǝw—ʔipah cɛp kǝliŋ.

13. Kayuʔ habis, neŋ tǝmpat heh bǝ-tǝdoh.

14. Boleh heh mǝsuarat. Mǝsuarat boleh heh pakat, pakat diʔ-leh. Heh gǝh dǝɲaʔ. Jaɲan heh rugiʔ dǝɲaʔ. Heh tahuʔ heh ciʔ, heh tahuʔ heh simpɛn.

15. Gob gǝn ʔiŋit kǝ-halbǝw gǝn boh kǝlapaʔ. Kiraʔ gǝn bunoh dǝɲaʔ. Kǝ-bah-lǝw heh gɔs. Jadiʔ gǝn bunoh dǝɲaʔ heh. Kǝ-sɛn heh ŋɔk sihat. Deʔ deh neŋ hɛhm sihat dah. Jadiʔ hukum heh samaʔ.

16. ʔipah haʔip masaʔ ʔaman. ʔiŋit, ʔipah haʔip. ʔipah pǝ-hinĕk.

17. Sudah kǝtis teʔ. ɲawaʔ tɔm—sǝkat. Pǝntiŋ bahayaʔ. Tɔm neŋ ʔum ʔayut. ʔoh lǝmpah. Teʔ lǝkɔc. *[Leads to]* teʔ runtoh. ʔoʔ diʔ ʔalor *[elsewhere]* ʔoʔ pǝcah ton.

18. Macam-macam, cariʔ makan—ʔuntoʔ makanan heh kayaʔ, tapiʔ dǝɲaʔ neŋ. Tahuʔ heh simpɛn. ʔayaŋ heh kayaʔ heh bunoh dǝɲaʔ. Ciɲhat heh ʔoʔ *[not]* bǝtet *[by being greedy for food.] [They take a little bit, keep it in reserve.]* ʔipah tahuʔ ʔipah simpɛn. Heh cariʔ makan, heh pǝntiŋ ɲawaʔ dǝɲaʔ. Tapiʔ neŋ gǝn tahuʔ pǝntiŋ ɲawaʔ dǝɲaʔ, ʔajoh dɛn.

19. *[We could be rich but]* pǝntiŋ ɲawaʔ dǝɲaʔ. *[Now they've changed a little bit but]* ʔipah ʔiŋit kǝ-sen. Sǝhidop ʔipah kǝ-hǝp, ʔipah bagiʔ ʔarahan. Jaɲan ʔipah bǝ-kǝlahi, bǝ-bunoh-bunoh.

20. *[There are Batek who want to be rich.]* ʔayaŋ sǝnaŋ—kǝsiksaan *[to live like this, but preferable to killing.]* Putusan ɲawaʔ gǝn diʔ.

21. *[We travel in one place then remember and go back there]*. *[People who live jinak, they]* bunoh dǝɲaʔ. *[JHEOA]* tinak ʔipah gɔs liar. Neŋ gǝn tahuʔ gǝn pikir. Yɛm pani gǝn tahuʔ caraʔ ʔakal.

Part II: Word-for-Word Translation

1. []-world-finish. []-world-finish. Already-no-tree.

2. When-they-dynamite/bomb, Gubar-long time-rain. [] we [EX]-remember.

3. []—do-only until-that. We [IN]-find-food-same. [] Negative marker-can-overly fond of-appropriately-that-we [IN]-find-food.

4. River-Tɔmoh-no-tree. Oil palm-only.

5. Soul-to live-upon-tree. []

6. []-island-[]-hold up-earth

7. That is past-we [IN]-peace-no-lose-world.

8. person-superhuman-to say: heart-earth-to make.

9. []-they-remember-they-miss.

10. To feel sorry for-song.

11. They-love [].

12. We [EX]-listen—what—we [EX]-hold-voice.

13. Tree-finish, no-place-we [IN]-to be shaded.

14. Can-we [IN]-meeting. meeting-can-we [IN]-decide together, decide together-do-then. We [IN]-refuse to give up-world. Don't-we [IN]-lose out on-world. We [IN]-know-we [IN]-eat, we [IN]-know-we [IN]-keep.

15. Malays-they-think-of road-they-put down-oil palm. Consider-they-kill-world. Where-we [IN]-live. So-they-kill-world-ours [IN]. Before-we [IN]-sit-healthy. Now-no-we want-healthy-already. So-law-us [IN]-same.

16. We [EX]-miss-time-peace. Remember, we [EX]-miss. We-show how.

17. Already-broken-earth. Soul-river—block. Important-danger. River-not-it wants-flow. It-spills. Land-soft. []-land-collapse. It-make-channel-[]-it-break-there.

18. Like-like, look for-food—for-food-we [IN]-rich, but-world-none. Know-we [IN]-keep. Not-we [IN]-rich-we [IN]-kill-world. Short-we [IN]-to become-[]-long-[]. [] We [EX]-know-we [EX]-keep. We [IN]-look-for-food, we [IN]-value-soul-world. But-not-they-know-value-soul-world, don't know-then.

19. []-value-soul-world. []-we [EX]-remember-before. As long as living-we [EX]-in-forest, we [EX]-give-instructions. Don't-we [EX]-fight, kill each other.

20. [] Not-easy—suffering []. End of-life-they-do.

21. [] []-kill-world. []-ridicule-we [EX]-live-wild. No-they-know-they-think. I want-them-know-way-reason.

Part III: Free Translation

1. Take away the forest, the world ends. We want people to know that the world can end. Already there aren't trees.

2. When they dynamite, Gubar makes it rain a long time. We [EX] remember.

3. Don't take away more forest, make that the limit. We [IN] look for food alike. We [EX] want to *pakat*. We [IN] cannot be overly covetous—we should take only what's appropriate for our livelihood.

4. Temoh River already has no trees. Only oil palm.

5. Our souls live upon the trees. The forest is the veins and tendons of our lives.

6. This land is an island. How can the land hold together without trees?

7. In the past, we [IN] lived in peace, we [IN] weren't losing the world.

8. The superhuman beings they say: "the heart of the earth they made."

9. The superhumans they remember they feel longing.

10. They feel sorry for us when they hear our [EX] songs.

11. They love us so they warn us what is happening.

12. We [EX] hear what they say, we hold on to their voices.

13. When the trees are gone, no place for us [IN] to shelter.

14. We [IN] can have a meeting. We meet, then we can discuss what to do. Discuss, decide, then we go ahead and do. Let's [IN] not give up the world. Let's [IN] not lose it. We [IN] should know how much to eat, how much to keep.

15. Malays in thinking of roads will lay down oil palm. Consider that they kill the world. Where is everyone [IN] going to live? So they kill our [IN] world. In the past, we [IN] lived healthy. Now, no longer can we [IN] want to be healthy. So everyone lives by common rules.

16. We [EX] miss the times of peace. We remember, we miss. We show how.

17. Already the earth is all cut up. The soul of the rivers is blocked. It's important to understand the danger. The rivers can no longer flow, they flood their banks. The soil becomes soft, it collapses. They open up channels elsewhere, that's where the earth fissures.

18. It's like, we [IN] look for food—from food we get rich, but the world is gone. We [IN] should know how to keep. It shouldn't be that we become rich and kill the world. Our [IN] lives are shortened, we don't live long, when we're too greedy. We [EX] know how to keep. When we [IN] make a living, we should value the soul of the world. But if they don't know to value the soul of the world, I don't know.

19. We [EX] could be rich but we value the soul of the world. Now we've changed a little bit but we [EX] still remember the past. As long as we live in the forest, we'll give the instructions. We [EX] don't want to fight, to kill each other.

20. There are Batek who want to be rich. It's not easy to live like this, we suffer. But this is preferable to killing the world, like life outside the forest. They bring about the end of lives the way they live.

21. We [EX] travel in one place then remember and go back there. People who live tame, they kill the world. JHEOA officers ridicule us for living wild. They don't know how to think. I want them to know how to reason.

Appendix B

Route Descriptions

To show how the landscape categories discussed in chapter 3 might be used concretely, in travel directions, I reproduce here two route descriptions composed by Lus (recorded in March 1996). She was in her early twenties. We were at Was Tabɛn (Malay Kuala Tabing) on the Tahaɲ River. I asked Lus to describe the land route from Was Yɔŋ (at the confluence of the Yɔŋ and Təmiliŋ rivers) to Was Tabɛn (see map 8). She said there were two: one direct route and the other somewhat more circuitous. After outlining the first route (somewhat perfunctorily) she warmed up and gave a more expressive account of the second route. We glimpse here a series of place-names, descriptions, and categories, that record the forest's ever-shifting qualities.

A cautionary note: what follows are recitations. They were not recorded during spontaneous conversation or recited just after a trip. As such, they are probably less detailed than if given in a natural setting.

Route 1

Galah Yɔŋ, lew kə-kəntəʔ Rənwis. Rilek (cwəh mih sar) balek nan Rənwis, ʔoh pibus lagiʔ tɔm naʔ Tənər. Wit-wit Tənər, lew Was Lamey, galah Lamey. Rilek Dayaŋ. Wit-wit Dayaŋ, lew kə-kaw—halbəw Bumbun.

Upriver on the Yong, reach the upper side of Rənwis River. Crest (ascend to descend) towards the Rənwis side where it meets the mother river Tənər. Go down the Tənər, reach Was Lamey, go up the Lamey. Crest over the Dayaŋ River. Down the Dayang, arrive here—the Bumbun[1] trail.

Route 2: Was Nuntoh Route that Bypasses Hayãʔ Tey

[Go to Was Nuntoh]. *Wit simpaŋ bah-kanan. Cip kə-udn. Halbəw bah-kiriʔ. Jəmraŋ ʔawãʔ tɔm. Cwəh kə-buloh ʔakar, rilɛk bah-dɔy, lew kə-kaw. Lew kə-ʔalor ʔawãʔ tɔm wɔŋ kə-kiyɔm bataŋ tɛras (bataŋ ʔuʔ sɔʔ manah jadiʔ tɛras). Cip diʔ kə-udn, sar bah-Tɔm Wal. Jəmraŋ gɛl udn. Cip cip cip cip deh udn, lew rənbak (kayuʔ ʔoʔ kiø, reʔ lagiʔ). Sar ʔalor. Cip cip deh kə-udn, lew lagiʔ kə-ʔalor, jəmraŋ ʔalor. Cip lagiʔ deh kə-wək lagiʔ ʔalor, jəmraŋ lagiʔ ʔalor. Cip. Lew simpaŋ bah-Tənər. Yey simpaŋ bah-kanan, pi-wel simpaŋ bah-kiriʔ. Cip nan was Tɔm ʔəmpat, lew təmpɛt [LTP] diʔ hayãʔ. Cip. Lew jəmraŋ Tɔm Wal. Cip lew kə-rənbak gil. Cip. Jəmraŋ*

ʔalor lagiʔ. Cwəh ʔalor. Sar lagiʔ kə-ʔalor. Cwəh ʔalor. Cip kə-Tɔm Hawap. [Lost fragment] *Tabar. Tətar bataŋ tɜras. Cip—lew kə- payaʔ, padaŋ ləbek. Cip. Jəmraŋ ʔalor lagiʔ. Cwəh ʔalor. Cip cip. Cwəh ʔalor lagiʔ. Jəmraŋ ʔalor. Cwəh bah-həp. Lew kə-ʔalor. Jəmraŋ ʔalor. Cwəh—blət bataŋ. Bə-liŋkut bataŋ udn. Cip. Sar bah-tɔm. Bah-tɔm—tɔm wɔŋ. Cip. Jəmraŋ ʔalor lagiʔ. Cwəh bah-həp. Cip cip lew kə-bataŋ. Tawɛs ʔoʔ kiø. Laŋkah bataŋ. Sar bah-tɔm. Jəmraŋ tɔm. Cip. Jəmraŋ tɔm lagiʔ. Lew kə-Hayaʔ Tɛy—təmpɛt heh lɔ̃t baniŋ.*

[Go to Was Nuntoh]. Downriver on the right turn-off. Walk there. [Take the] left path. Cross the stream. Ascend towards the *buloh akar* [bamboo; unidentified], then cross over in this direction, to come here. Make for a small stream channel that is under a *tɜras* (the trunk's rotted, long been a *tɜras*). Walk there, descend to Wal River. Cross that side of the river. Walk walk walk walk on, arrive at the *rənbak* (newly fallen tree). Go down to the stream channel. Walk walk on from there, walk on where there's yet another stream channel, cross that one. Walk on where there's yet another stream channel, cross once more. Walk. Arrive at the turn-off to the Tənər. Ignore the right turn, turn off to the left. Walk from the mouth of Tɔm ʔəmpat, arrive at the place where you made your lean-to.[2] Walk. Arrive at the point of crossing for Wal River. Walk until you get to the *gil* (*Koompassia excelsa*) tree *rənbak*. Walk. Cross another stream channel. Walk up from that. Descend into another channel. Ascend. Walk to Hawap Stream. . . . *Tabar* palm (*Arenga obtusifolia*). Cross the *tɜras* log bridge. Walk—arrive at a swamp, a single-species grove of *ləbek* (unidentified plant). Walk. Cross the stream channel again. Go up from it. Walk walk. Go up again. Cross the stream channel. Ascend into the forest [away from the watercourse]. Arrive at a stream channel. Cross the channel. Climb up—walk under a tree. Walk round that tree. Walk. Go down to the river. At the river—a side stream. Walk. Cross the stream channel. Ascend into the forest. Walk walk until you arrive at a tree trunk. A fallen *tawɛs* tree. Step over. Go down to the river. Cross the river. Walk. Cross the river again. Arrive at Hayã̆ʔ Tɛy—the place where we ate *baniŋ* (fruit of the Great Spindle Ginger; *Hornstedtia scyphifera* var. *grandis*).[3]

Notes

1. Trail to the Bumbun Yong [Yong Hide of Taman Negara].
2. Refers to a rough shelter that I had put up a month earlier: I was alone, an hour short of reaching camp, when caught in a torrential downpour. For days afterwards, users of the trail told me that they had sighted my lean-to.
3. Refers to a camp-moving day in September 1995 (i.e., six months earlier). En route to the new camp, we made a number of stops to harvest fruits, among which was *baniŋ*.

Appendix C

Notes on Wild Yam Species

This appendix brings together my notes on wild yam species, most of which are derived from Batek comments. Idiosyncratic comments are identified by informant names. Batek ratings about each species are also provided. In addition, the following names were collected in free-listing situations, but without supplementary notes: *?obɛt cɛ? (Stemona tuberosa), bəsu?, carga?, dago?, həw, maŋkel, mantoh, paŋɔn, pumaŋ, purɔn* (identified as *Schoenoplectus articulatus*), and *taloŋ*.

Abbreviations: D: Dioscorea. HD: hard to dig for. ED: easiest to dig for. GF: good flavor. NF: no or poor flavor. C: common

1. **ciŋ?il** *(D. alata)*
Notes/comments: grows near Malay dwellings; banks of the main rivers, in flat areas, secondary forest. ?eyNɔn: also found in Kelantan
Rated: GF (ta?Kadɔy)

2. **cariŋ**
Note: observed growing in a hilly environment

3. **cərgal**
Note: ?eyNɔn: found in Kelantan

4. **ciyak**
Notes/comments: Fish poison can be extracted from the tuber. Thorny. Distinguished from a subspecies or variety, **ciyak manroŋ** [lit. skink *ciyak*], which has no thorns. ?eyNɔn: also found in Kelantan

5. **cway** (some identify its avoidance name as **tampa?**).
Notes/comments: ta?Kadɔy: yield is erratic: if you get find the fleshy tubers one day, on the next you'll get none. ?eyNɔn: also found in Kelantan.

6. **gadoŋ** *(D. hispida)*. Batek name **kədat**
Notes/comments: known as a common famine food for rural peoples in Asia and one of the most commonly harvested species in Endicott's field site. Never harvested during my field studies. It contains the alkaloid *dioscorine*, that must be leached out before it's safe to eat. Batek women identify this (the processing difficulty) as the reason why they shun it.

7. jabet
Note: grows near Malay dwellings

8. hakay (Elephant-foot yam; *Amorphophallus campanulatus*)
Note: grows near Malay dwellings

9. kasi?
Note: grows near Malay dwellings

10. kəbak
Notes/comments (na?Caŋkãy; na?Gk): grows in *te? laŋeh* (hill environments), which is its place of origin. Smelly.

11. kənsey
Note: grows on slopes and *te? laŋeh* (hilly environments)
Rated: GF (ta?Kadɔy)

12. kətit (*Smilax* sp.)
Notes: known to contain medicinal properties; food value not determined

13. pam
Notes/comments: the root is very long. Kayə? says that if it *lis cəriboŋ* (grows slantwise down) there'll be fleshy tubers. And if it *ciŋ* (grows horizontally) then it could go on for many feet. Tubers usually quite fleshy. Growth and habitat observations: grows land extensively; near rivers on flat ground. Na?Alɔr: also near Malay dwellings; by the side of logging roads, near Malay rubber trees.
Rated: HD (Kayə?; na?Alɔr [relative to *takop*]); GF (ta?Kadɔy)

14. patiy. Avoidance names **təruy; tə-miyaŋ** (that irritates the skin)
Notes/comments: many taboos related to its processing. Leaves cause skin irritation. Rarely harvested: yield is dismal. Long root; rarely produces tubers. Tubers tend to be small and flesh-poor. Widely distributed along the Tənər (Malay Tenor). Grows in clusters. ?eyNɔn: also found in Kelantan
Rated: HD (na?Ksɔ?; ta? Kadɔy)

15. payol
Note: grows in *te? laŋeh* (hilly environments)

16. rɛm (*D. prainiana*). Avoidance name **las.**
Notes/comments: the enormous root grows in one spot near to surface and doesn't spread in every direction. Three varieties (subspecies?) distinguished by different colored flesh: red/vermilion; blue/light purple; dirt-like brown. Some varieties are edible raw. ?eyNɔn: also found in Kelantan. Beliefs and mythology associate it with the tiger.
Rated: ED (Kayə?; ?eyGk)

17. rɔ̃n
Note/comment: causes skin irritation.

18. tahoŋ
Notes/comments: tubers are long. Grows close to ground surface; on banks of major rivers and tributaries; also near Malay dwellings. ʔeyNɔn: also found in Kelantan. One variety or subspecies identified: *tahoŋ roh*.
Rated: GF (taʔKadɔy); ED (naʔGk)

19. takop (*D. orbiculata*)
Notes/comments: grows deep in the ground; the root travels widely and extensively over a large area. Has been spotted on the verge of logging roads. NaʔAlɔr: also near Malay dwellings; by the side of logging roads, near Malay rubber trees. NaʔCaŋkãy; in *teʔ laŋeh* (especially big mountains). TaʔJamal: a dependable food source; during the rainy season when looking for a place to set up camp, they also check to see if the location contains *takop*. NaʔKapey identified a variety or subspecies, *takop saroʔ* (lit. ghost *takop*), which is extremely hard and woody. It is said that these were put aside by ghosts for their own sustenance. Another variety (?), *takop wak*
Rated: NF (taʔKadɔy); C (taʔRaŋlɔs); ED (naʔAlɔr [relative to *pam* and *wɔh*]); HD (ʔeyGk)

20. tampak
Notes/comments: grows close to surface in small bunches coming off the vine; woody tubers. Observed growing in flat, moderately sloping ground, advanced secondary growth.
Rated: ED (general)

21. tãw
Notes/comments: identified by Kirk Endicott as a *teʔ laŋeh* tuber. But observed in flat, moderately sloping ground as well.

22. wɔh
Notes/comments: grows land extensively. Kayɔʔ: produces many flowers that drop close together; thus the vines tend to be found in clusters.
Rated: HD (naʔAlɔr [relative to *takop*])

Glossary

This glossary lists the majority of Batek terms and particles that appear in the book. It includes noteworthy place-names (toponyms) and river-names (hydronyms), most of which are cross-referenced to the maps in this book. Specialized words listed elsewhere (for example, in tables and the wild yam names in appendix C) are not presented here.

The Batek language, like that of many other Orang Asli peoples, is heavily influenced by Malay, the majority language in Malaysia, and this is reflected in a large number of loanwords. For these, the Malay spelling (=M.) follows the Batek headword in parenthesis. Where the Batek meaning differs slightly from the standard Malay definition, the latter is also provided.

In Batek, hydronyms are formally preceded by the word "Tɔm" (meaning "river"); for example, the Kechau River (Malay Sg. Kechau) would be Tɔm Kəciw in Batek. Here, only the unique names are given—in Batek and, wherever relevant, Malay. Thus, the Kechau River appears here as "Kəciw (M. Kechau)."

bab food or starchy food (generic).

bah- at, to; preposition indicating directionality. Most commonly prefixed to locative nouns (*bah-kəntə?* [upriver]; *bah-kiyɔm* [below; downriver]; *bah-kiri?* [to the left]; *bah-həp* [to the forest]; *bah-te?* [groundward]).

bakar (M. *bakar*) to burn (vegetation only).

banar (M. *bandar*) town, city. Less commonly used than *dəŋ. ?oraŋ banar* or **batεk banar** urban peoples.

banyir (M. *banjir*) flood.

baŋkol (M. *gaharu* or *kayu wangi*) eaglewood (*Aquilaria* spp.).

baŋkɔŋ (M. *bangkong*) wild jackfruit (*Artocarpus integer* var. *silvestris* Corner).

baŋsa? (M. *bangsa*) race or ethnic group; type or group of similar objects.

bat to land or perch on.

batεk people. **batεk həp** people of the forest.

batu? (M. *batu*) stone or rock. **batu? cɔnεl** mythicized stone or rock: i.e., rock

formations whose origins are explained in etiological myths. I had not heard the word *cɔnεl* as an independent noun. However, in Jahai language *cɔnεl* suggests mythic stories (Niclas Burenhult 2002, personal communication).

Batu? Səmbilan (M. Batu Sembilan) Malay village on the edges of Kenong State Park, Pahang. The Batek maintain a semi-permanent camp nearby.

bawac pig-tailed macaque (*Macaca nemestrinus*).

bay to dig. **bay takop** to dig for *takop* tubers; the process of collecting wild tubers in general; living off wild as opposed to cultivated plants.

ba? question marker.

bε̃c predator (generic); secondary name of the tiger.

bəlaw blowpipe.

Bəlaw (M. Blau) a tributary of the Təmiliŋ River in Taman Negara. Named for *bəlaw*, the Batek word for "blowpipe." Blau Jetty is the boat jetty located at the confluence of the Bəlaw and the Təmiliŋ. [Map 8].

bəla? singly. **yɛ? bəla?** on my own.

bəldɛ̃l broad path.

bəlhɔt watery texture (of tubers). **bəlahɔt** term of abuse for small-sized *takop* tubers.

bəlukɛr (M. *belukar*) secondary forest.

bənəm very tall mountain. From the classic Mon-Khmer word for "mountain"; cognates include Chewong *bənim* and the "Phnom" in Phnom Penh.

bəplit longer route, i.e., not a shortcut.

bərguh to be successful in any harvest of food.

bətow straight (of river or trail).

bəw big. **tə-bəw** that is big.

bit-bɛt lit. to hang on tightly here and there: i.e., to be entwined, derived from *bɛ̃t* (to hang on tightly).

boh to put down.

bom (English *bomb*) to bomb or set off any explosives.

boŋa? (M. *bunga*) flowers (generic). Those of yam species are called *bəkaw*.

bɔt to take.

bukit (M. *bukit*) hill.

buloh (M. *buluh*) bamboo (generic).

Buməkəl affluent of Tɔm Təkɛl in Kecau sub-district. The Batek maintained a camp on the upper reaches of this river. [Map 7].

bumutlit to meander (of paths and rivers).

bunoh (M. *bunuh*) to kill. Rarely used; in commonplace talk, this has not supplanted Batek *sakɛl*.

cam (M. *cam* [recognition by sight]) to search for. **cam bab** to look for food; to undertake any type of foraging activity.

can foot. **can banyir** lit. foot of the flood: i.e., onset of the flood season.

cara? (M. *cara*) way of doing things.

cawas *Artocarpus lanceifolia*. Culturally one of the most important fruits.

cɛp to hold or grasp.

cəba? hillock or ridge-crest.

cəmcəm *Calamus castaneus*. The principal source of lean-to thatch materials.

cəŋrəŋ clear (water/river).

cəpik bearcat (*Arctictis binturong*).

ci? to eat (generic); to eat starchy foods.

cip to walk; to go. **cip bah-həp** to go to the forest.

cirloŋ to lose the way.

cwəh to ascend (vertically and geographically). **cwəh bah-həp** lit. to go up to the forest: i.e., to enter it. **(halbəw) cəniwəh** path of ascent.

Dada? Kriŋ (M. Kg. Dada Kering) Malay village located at the southwestern border of Taman Negara, between the highway to Gua Musang and the Batek Tanum village of Marɛm. [Maps 3 and 7].

Dayaŋ (M. Dayang) tributary of the Tahaɲ River in Taman Negara.

dəɲa? (M. *dunia*) world.

dəŋ house; village; town. From the Proto Mon-Khmer word for "house on stilts."

di? to do or make something.

dok poison (especially blowpipe poison from the ipoh tree *Antiaris toxicaria*).

dut avoidance name of the pig; the type of nose associated with the pig; the inedible stub of edible roots for *pam* tuber.

gadoŋ *Dioscorea hispida*. Also known by the Batek name *kədat*, which is memorialized in the story of Pa? ?aŋkol.

gajah (M. *gajah*) the most common name for the elephant (*Elephas maximus indicus*), whose other names include *gawəy* and *gago?*.

gəh to refuse to give.

gən they (third person plural pronoun).

git?ac wet and soggy ground.

gil *Koompassia excelsa*. A common bee-tree, the tallest legume in Southeast Asia, and sometimes listed among the original trees.

gob outsider, stranger—prototypically Malays. **gob ?amɔ?** lit. Malays went amok: i.e., slave-raiding. **masɔ? gob**

lit. entering Malayness: i.e., to convert to Islam.

gɔs to live; to be alive. For various philosophical reasons, I prefer the gloss "to dwell" for this word.

Gubar thunder; the thunder-god. Though regarded by analysts as a "god," Gubar really has a double persona: he is both an other-worldly entity whose anger can destroy the world and a Trickster-like buffoon who is subject to verbal insults and mockery.

gul slow-moving water.

gunoŋ (M. *gunung*) mountain.

Gunoŋ Tahaɲ (M. Gunung Tahan) tallest mountain in the Peninsula, a popular climbing destination in Taman Negara. [Maps 4 and 5].

hal spoor, track, footprint. **kes hal** lit. to track; metaphorically, to follow.

hala? shaman. **hala? ?asal** superhuman being.

halbǝw (=harbǝw) path; route (i.e., "pathway"). Burenhult has uncovered a parallel term from the Jahai—for them, the *har* is a small path (Burenhult 2002, 265). If this etymological root is proved to hold true for the Batek as well, *halbǝw* and *harbǝw* should be compound words, formed from combining the words for "tracks or footprints" (*hal*) and "path" (*har*) to the adjective "big" (*bǝw*). Following that logic, *halbǝw* would have the literal meaning of "big tracks" and *harbǝw* "big path." **halbǝw banar** urban road. **halbǝw niwaŋ** path approaching a branching point. **halbǝw pǝlancoŋ** tourist trail. **halbǝw tah** surfaced road. **halbǝw wiŋ** logging road.

haŋep night.

haɲiw quiet (of atmosphere).

hapoy lean-to thatch.

Hariŋ (M. Aring) an important tributary of the Lebir River in Kelantan and one of the major territorial markers there. Rises from the northern flanks of the Tahan range. [Map 4].

hayã? lean-to; by extension, the camp. From the Proto Mon-Khmer word for "house."

Hayã? Tey lit. 'Tey's Lean-to.' Location of a Batek camp on the upper reaches of the Wal River in Taman Negara. Named for a Chinese man ("Mr. Tey") who once stayed there. [Map 8].

ha?ip emotions associated with longing and nostalgia.

hãw to eat vegetable-like foods.

heh you-all and we-all (first person plural inclusive pronoun).

hǝŋ?u? overgrown (of vegetation).

hǝp forest. **hǝp lǝy** standard lowland forest.

hõnril the strata of a riverbank wall that appears when the water level goes down.

hinĕk to imitate. **pǝ-hinĕk** lit. to cause to imitate: i.e., to show how something is done or said.

hnadaŋ ridge-top where slopes meet; may be in the form of a long plateau. Considered the natural boundary of a watershed.

hukum (M. *hukum* [order, command, judicial sentence, legal rules]) laws.

jaga? (M. *jaga*) to guard; to be astir. **jaga? hǝp** to look after the forest.

jǝbec useless, bad.

jǝlew long-tailed macaque (*Macaca fascicularis*).

jǝm cold.

jǝmaga? unmarried male—a category usually associated with youthfulness and vigor, but also includes older men, like widowers and divorcees; handsome.

jǝniŋwaŋ any kind of branching point, be it of overland trails, rivers, tree branches, fingers, and toes. Possibly a derivation of *niwaŋ*.

Jǝrantut (M. Jerantut) capital of Jerantut district. [Map 3].

jina? (M. *jinak*) tame (of animals).

jok to travel from one place to another. Contrasted to *lǝp* (to travel for a specific purpose, like a collecting expedition lasting a few days, and return to the point of origin).

kabɛn friend; kin.

kapal bəsi? (M. *kapal besi*) lit. iron ship: i.e., airplane.

kawaw bird (generic term). **kawaw ?asal** original birds. **kawaw hala?** superhuman birds. **kawaw hantu?** ghost birds. **kawaw tahun** fruit season bird.

kaya? (M. *kaya*) rich.

kayu? (M. *kayu* [wood]) general term for all trees.

kã?kɔ̃? to say *?ɔ̃h*.

Keladong Camp tourist campsite in Taman Negara named for a Batek headman.

kə-sɛn that is past. **heh kə-sɛn** we all in the past. **cip kə-sɛn** to walk in front of a line.

kəbi? general term for all fruits.

kəbon white-handed gibbon (*Hylobates lar*).

Kəciw (M. *Kechau*) forms the western boundary of the territory in Pahang and after which the sub-district of Kecau is named. Its source is in the Tahan mountain range. One of the traditional routes to Pahang from Kelantan follows this river's topographic logic. [Maps 3, 4, 5, and 7].

kədah unmarried female—a category usually associated with youthfulness and vigor, but also includes older women, like widows and divorcees; beautiful. Plural *kəradah*.

kədidi? sandpiper bird, probably *Actitis hypoleucos*.

kəjeŋ to hear. **kenjeŋ** to notice.

kəlamin (M. *kelamin* 'pair') a childless couple.

kəlaŋes heart.

kəliŋ general term for any sound, noise, voice, or language.

kəliŋwəŋ crooked (generic).

kəlpah daytime.

kəmam a family consisting of parents and children.

kəmɔyɛn (M. *kemian*) *Styrax benzoin*. Produces a resin used in curing and religious rituals.

kənmoh any name or label. **kənmoh ?awã?** or **kənmoh kancok** lit. child's name or grandchild's name: i.e., teknonym. **kənmoh pənəwak** lit. name of disguise: i.e., nickname or avoidance name. **kənmoh tinak** term of abuse.

kəpok orphan. **bə-kəpok** lit. to be an orphan; metaphorically, living alone away from parents, kin, or friends.

kərja? (M. *kerja*) to work. **kərja? ?awey** to "work" (collect) rattan.

kit?ɔ̃t to circle round an area (of tigers).

kipəy to fly.

kiyɔm locative noun: under, underside, below.

kiy-kuy an orientation term meaning 'to be heading'—a reduplicative verbalization of the noun *kuy* (head). For river flows, *kiy-kuy* signals where two contiguous rivers meet ("head towards") each other. In a sleeping position, it would signal where the head is pointed towards.

klapa? (M. *kelapa sawit*) oil palm. Malay *kelapa* refers to the coconut, which is *ɲior* in Batek.

klet unripe (of fruits).

komini Communists.

Kɔh (M. *Koh*) river in Kelantan associated with the Batek Te' sub-group. *Kɔh* is the name of an unidentified species of mud-turtle. [Map 3].

kraja?an (M. *kerajaan*) government.

kunah to round and round (of rivers). Cf. Jahai *kunar* (river bend). Most likely derived from English *corner* via Malay *kuna*.

kwoŋkwɔc owl.

labi? large river turtle. Features in some creation myths.

Lal location of a Batek camp on the eastern banks of the Təmiliŋ River. [Map 8].

lalaŋ (M. *lalang*) grassland.

lawac a class of prohibitions associated with the thunder-god.

lawak to look for ground-level foods (general term).

lew to leave; to arrive at. **lew kə-kəntəʔ** to arrive at the upper side of.

Ləbər (M. Lebir) probably the major Batek territorial marker in Kelantan. It becomes the Kelantan River at the lower reaches. Name derived from a Mon word meaning "sea." [Maps 3 to 5].

ləkɔc soft to the touch, pliable.

Ləpan Dok lit. 'Flat-land of Ipoh Trees.' Location of a Batek campsite off the Tirpal River in Taman Negara. [Map 8].

ləspəs ʔuʔ ral the end of the rainy season.

ləwey honey-bee.

ləʔ indicator (species).

lə̃t to eat fruits.

linaŋ to walk over the crest of a ridge/hill/ mountain. I reserve the possibility that the Batek word is a compound from *lih* (body) and *naŋ* (side of or perpendicular to).

lis the growth of a tuber.

lõʔ to wash into the river (of earth).

-m intention marker. **Yɛm** I intend to. **Mohm** you intend to. **Gəm** they intend to. **Hehm** we-all and you-all intend to. **ʔohm** she/he/it intends to.

malaŋ (M. *malang* [unlucky]) to be crippled by ill-luck in any harvest of food.

malɛs a general purpose adjective suggesting displeasure or reluctance.

manah long (of duration). **mənanah** to be a long time. **hayã̃ʔ manah** old camp.

Marɛm (M. Kg. Sg. Garam) Batek Tanum village near the Malay village of Dada Kering. [Map 7].

masaʔ (M. *masa* [time]) time in general, but commonly refers to a period (i.e., a stretch of time). **masaʔ banyir** rainy season. **masaʔ tahun** or **masaʔ kəbiʔ** fruit season. **masaʔ ʔaman** time of peace.

mɛ̃t lit. eye; metaphorically any core part of an object, like the seeds of fruits. **mɛ̃t kətoʔ** sun. **mɛ̃t ʔam** nipple.

mənantaŋ (M. *binatang*) animal (generic).

məsuarat (M. *mesyuarat*) to have a formal meeting (rather than a casual get-together).

moh you (second person singular pronoun).

mohʔɔŋ source of a river. For Jahai, *ʔˀŋ* is a secondary word for "water" or "fluid" so for them *mɔh ʔˀŋ* literally translates as "nose of river"; i.e., its source (Niclas Burenhult 2002, personal communication).

mɔs end (an old Austronesian, but not Malay, word). **mɔs tɔm** lit. end of the river: i.e., upper reaches.

naʔ mother.

ɲawaʔ (M. *nyawa*) soul.

ɲoʔ to lie; to deceive; to err.

ŋɔk to reside in; to sit down. **ŋɔk jəmit** to sit still; to stay put in an area rather than to move to other locations. I am unsure which of these meanings is a metaphoric extension of the other.

pacat (M. *pacat*) land leech. Alternate name *klɔm*, from which is derived the verb *glitɔm* (to arch; i.e., to curl like the body of a leech).

padaŋ (M. *padang* [field, playing ground; originally treeless wasteland]) single species grove.

padiʔ (M. *padi*) rice.

pakat (M. *pakat*) to decide what to do together.

papar slope; small ridge.

paw side. **paw halbəw** trailside or verge. **paw-paw** to walk alongside. Cf. Jahai *paw* (side of the chest from the armpit down to the bottom of the ribcage).

pawɛs to be unsuccessful in any harvest of food.

Payaʔ Kladiʔ (M. Paya Keladi) Batek Tanum village.

paʔ father.

pə- causative marker. How its functions differ from *pi-* is not yet clear.

pə-kritis lit. to cause to break apart severally: i.e., to cut up (meat) for distribution. Derived from *kətis* (break apart, sever).

pənrəʔ to dig at a previously harvested yam root, because it has continued to produce tubers.

pərban generic category for forest clearings opened up by treefalls.

pərəŋdəŋ broad path.

pərkac muddy (water/river). May be a causative: "to cause one to scratch."

pi- causative marker. **pi-dɛŋ** lit. to cause to see: i.e., to point. **pi-wəʔ** to cause to rise up (of human body); to cut off the fleshy tubers of *wɔh*, *pam* and *kənsey*. **pi-wel** to make a turn-off.

pibus to meet (of rivers).

ploʔ edible fruit (generic).

polaw (M. *pulau*) island.

pumpakoh Indian cuckoo (*Cuculus micropterus*).

radiʔ to be overly fond—and possibly covetous—of something (principally, food). Malay *radi* (of Arabic origin) means "may God bestow favor"; whether this is the root of Batek *radiʔ* is not clear. The meanings are certainly not similar.

Ralat a tributary system on the middle reaches of the Kechau River. Some of its affluents flow directly down from Taman Negara hills. Ralat is the main trunk of this system. [Map 7].

ray lightning scar in a forest patch.

rel a still-standing but dying tree.

rəmram piled up driftwood.

rənam fragrant leaves collected for bodily ornamentation.

rənbak any newly fallen tree that has not begun to rot.

rənɛm to set up a fishing rod (i.e., to leave the rod standing while waiting for fish to bite).

rɛɲ to eat meat.

rilɛk to walk over the crest of a ridge/hill/mountain.

ripat early evening, just before nightfall.

rugiʔ (M. *rugi*) any loss or non-physical injury.

runtoh (M. *runtuh*) to collapse.

Ruwiw (M. Ruil) a Taman Negara tributary of the Tembeling River, southwest of Was Yɔŋ. An extensive river system, fed by many small affluents. Soil has a heavy limestone component. [Map 8].

sakey (M. *sakai*) the Malay word *sakai* is a term of abuse commonly leveled at Orang Asli; has a range of meanings like "slave" or "dependent." The Batek word *sakey* refers to a kind of cannibal that eats raw meat and is sometimes used to insult naughty or unruly children. In the colonial era "Sakai" was used by writers and administrators as a general term for Orang Asli or sub-groups among them. **Sakai Liar** Wild Sakai. **Sakai Jina'** Tame Sakai.

sakɛl to hit or to kill.

saɲyɛt to be uninhabited (of place).

sar to descend (vertically and geographically). **sar bah-dəŋ** lit. to go down to the village/town. **(halbəw) pənisar** path of descent.

sayɛŋ (M. *sayang* [to love]) to love or be fond of—and implicitly not want to lose—a person or thing.

sec meat (generic).

sɛn past (in time); ahead (in space).

səkat (M. *sekat*) to block or be blocked.

sətsɛt spiderhunter bird (*Arachnothera longirostra*).

sə̃t beeswax.

siyal to fail to find animal tracks or tuber vine.

sɔʔ to rot; any rotten matter (commonly, corpses and carcasses).

surat (M. *surat*) letters, books, writing. **surat kitab** from the Malay for "sacred paper." Not ever heard in conversation outside the context of the *bakar lalaŋ* myth of origin.

tabiŋ to swing from branch to branch with the arms. Contrasted to *plibat*, which is to achieve the same effect with the full body.

tahan (M. *tahan*) to withstand.

tahat to hear; to see.

Tahaɲ (M. Tahan) a tributary of the Təmiliŋ. One of the major territorial markers in Taman Negara. [Maps 3, 4, 5, and 8].

tahun (M. *tahun* [year]) seasonal fruits; year. **tahun bəw** lit. big fruits or big year: i.e., a fruit season of abundance. **tahun cəkey** a lean fruit season.

tahu? (M. *tahu*) to know.

takop general term for all wild yams, from the name of a widely distributed species, *Dioscorea orbiculata*.

tala? to flee or to escape.

talo? dusky leaf monkey (*Presbytis obscura*).

Taman Ngara? (M. Taman Negara) the oldest national park in Malaysia, most of which straddles Batek territory in the three states. The largest remaining tract of unbroken forest left to them. [Maps 3, 5, 7, and 8].

tanyoŋ (M. *tanjung*) any land type that juts out like a peninsula, such as the capes of rivers or promontories.

taŋkap (M. *tangkap* [to catch]) to entrap.

tasek ripe (of fruits).

tasoŋ to be lazy to go to the forest.

tawăk butterfly.

ta? grandfather; old man.

ta?a? palm piths and cabbages (generic).

taɲol to pine away from feeling *ha?ip*.

tek to sleep.

te? land; all the land (that is, "the earth"); soil (the physical stuff); ground.

te? baros limestone soil forest.

te? laɲeh mountain or high ridge forest.

te? təraɲ forest located on hilly slopes.

te? tom hal lit. land water tracks: i.e., widely traveled (an idiomatic expression, roughly similar to "he's left his footprints everywhere").

tə- relative marker. **tə-bɔ̆ŋ** lit. that starts or starter: i.e., trailhead.

təbeŋ rise in a path.

təbiŋ tom (M. *tebing sungai*) riverside.

təlo? bəw (M. *teluk besar*) big bend of a river.

Təlo? Gunoŋ (M. Telok Gunong) Batek Tanum village in Cegar Perah on the Temetong River near Jerantut town.

Təmaw (M. Temoh) one of the long-flowing rivers bridging the Kecau and Tembeling sub-districts.

təmbuah deep waters.

Təmiliŋ (M. Tembeling) a tributary of the Pahang River and itself the main trunk of an extensive drainage area; meets the Jelai River at Kuala Tembeling. Forms the traditional boundary between the Batek and Semaq Beri territories. Also the southeasternmost edge of Batek territory and the southeastern border of Taman Negara. An important waterway and trade route in early Peninsula history. [Maps 3, 4, 5, and 8].

təmkal man; male. Plural *təmakal*.

təmpet (M. *tempat*) place.

Tənər (M. Tenor) a long-flowing tributary of the Tahaɲ River in Taman Negara, on the boundary between Lipis and Jerantut districts. [Map 8].

təras a dead, long-rotted tree.

tərɛn to hunt from the tree-top (i.e., rather than shooting from the ground-up).

tətər to walk on a switchback or zigzagging trail. **halbəw pənitər** trail with switchbacks.

tinak to ridicule.

tiwəh to walk around obstructions. **(halbəw) pəniwəh** path around obstructions.

Tohan (M. *Tuhan* [God]) one of the named creator beings. He is not talked or appealed to like an ever-living presence. Neither beneficent nor malevolent, he has no practical role in human affairs. He seems to exist only for the function marked in the creation stories and nothing else.

tom river; water. **?awă? tom wɔŋ** stream affluent. **tom kəmaraw** dry season river. **tom liwin** waters turning slowly round and round. **tom na?** main river or trunk river. **tom wɔŋ** tributary. **tom**

zĩl flood season river, when it floods its banks.

toŋ masked palm civet (*Paguma larvata*).

tɔt to see.

Treŋin (M. Trenggan) there are two rivers of this name in Batek territory: the better known Trenggan River in Taman Negara that drains into the Təmiliŋ [Map 8]; and another one in the Atok area flowing through the Yung Forest Reserve.

trichŏc very narrow path.

tuha? not quite ripe (of fruits). Many are edible already at this point.

tupan forest clearing dominated by big-sized fallen trees.

Wal (M. Wa) a tributary of the Treŋin River in Taman Negara. [Map 8].

was confluence or mouth of rivers (noun); to split (verb).

Was Lipis (M. Kuala Lipis) capital of Lipis district. [Map 3].

Was Tahaɲ (M. Kuala Tahan) located at the confluence of the Tahaɲ and Təmiliŋ Rivers. There are two parts: that on the western bank of the Təmiliŋ is the headquarters of the national park Taman Negara; the eastern part is a Malay village. [Maps 5 and 8].

Was Tirpal location of a Batek camp set beside one of the tributaries of the Tənər River in Taman Negara. [Map 8].

Was Yɔŋ (M. Kuala Yong) semi-permanent settlement in Taman Negara, about a ten-minute boat ride downriver from Was Tahaɲ. Located at the confluence of the Yɔŋ and Təmiliŋ Rivers. Spontaneously established by the Batek after they were told to leave their original settlement in the Was Tahaɲ park headquarters area. [Maps 3, 5, and 8].

Was ?ato? (M. Kuala Atok) settlement on the boundary of Taman Negara; administered by JHEOA. Located at the confluence of the ?ato? and Təmiliŋ Rivers in Taman Negara. About an hour's boat ride downriver from Was Yɔŋ. [Map 3].

wek to return.

wĕc any pass or gap between hills and mountains.

Wĕc Sok Bawac lit. 'Pig-Tailed Macaque Fur Mountain Pass.' Located at the mouth of a tributary of the Ruwiw system in Taman Negara. [Map 8].

wɔŋ a common Orang Asli word for "child"; it does not appear in Batek as such but is the root for *mawɔŋ* (all female mammals, including humans, in the nursing stage of life). tɔm wɔŋ lit. child river: i.e., tributary of a river. bolan mawɔŋ a full moon.

yah tiger (*Panthera tigris corbetti*); secondary names ?ayŏ?, bĕc. Terms of abuse include Cikok Te? (Earth Digger) and Ha?ac Kalkok (Smelly Claws).

yalɨw woman; female.

ya? grandmother; old woman.

yɛ? I (first person singular pronoun).

?ajoh don't know. ?ajoh dɛn dɛn may be an emphatic marker; it adds a note of resignation.

?alor (M. *alur*) stream channel or water channel.

?asal (M. *asal*) origin; original.

?awă? child; offspring; diminutive marker. ?awĕ? term of endearment for one's own children (particularly preadolescents), or, less commonly, any younger person with who one is on familiar terms.

?awey woody and non-woody vines, including rattans (generic).

?ay all nonhuman animates (generic); numerical classifier for animals. ?ay Həp (The Forest Animal), ?ay tə-Bəw (The Big Animal), ?ay Cəribuŋ Lenti? (The Slanting Tongue Animal): avoidance names of the tiger, elephant, and Giant Malaysian Frog, respectively.

?ayaŋ no.

?ey father. Used only in the teknonyms.

ʔəntiŋ afraid.

ʔɔ̃h "goodbye."

ʔikan (M. *ikan*) fish (generic).

ʔikut kraja ʔan (M. *ikut kerajaan* 'follow the government') adopt government-approved norms.

ʔiŋal noise.

ʔiŋit (M. *ingat*) to remember.

ʔipah we but not you (first person plural exclusive pronoun).

ʔobɛt (M. *ubat*) medicine.

ʔoh she, he, it (third person singular pronoun).

ʔolar (M. *ular*) snake (generic).

ʔɔt to hide from.

ʔujan (M. *hujan*) rain. Also *hac*.

ʔurɛt (M. *urat*) veins and tendons.

ʔuylah term of address for an intimate (close friend, spouse, or kin).

References

Abdul Rahim Nik. 2001. "Forestry." In *Malaysia: National response strategies to climate change*, edited by Chong Ah Look and Philip Mathews, pp. 303–37. Kuala Lumpur: Ministry of Science, Technology and the Environment, Malaysia.

Aiken, S. Robert. 1973. Images of nature in Swettenham's early writings: A prolegomenon to a historical perspective on Peninsular Malaysia's ecological problems. *Asian Studies* 11:135–49.

Anonymous. 2002. "Owl expert's 'Harry Potter warning,'" *BBC Online*, 11 May 2002, <http://news.bbc.co.uk/hi/english/uk/wales/newsid_1981000/1981651.stm>.

Benjamin, Geoffrey. 1973. "Introduction." In *Among the forest dwarfs of Malaya*, by Paul Schebesta (reprint of 1928 ed.), pp. v–xiv. London: Oxford University Press.

———. 1976. "Austroasiatic subgroupings and prehistory in the Malay Peninsula." In *Austroasiatic studies, part I*, edited by Philip N. Jenner, Laurence C. Thompson, and Stanley Starosta, pp. 37–128. Honolulu: University Press of Hawaii.

———. 1997. "Issues in the ethnohistory of Pahang." In *Pembangunan arkeologi pelancongan negeri Pahang*, edited by Nik Hassan Shuhaimi bin Nik Abdul Rahman, Mohamed Mokhtar Abu Bakar, Ahmad Hakimi Khairuddin, and Jazamuddin Baharuddin, pp. 82–121. Pekan, Malaysia: Muzium Pahang.

———. 2002. "On being tribal in the Malay world." In *Tribal communities in the Malay world: Historical, social and cultural perspectives*, edited by Geoffrey Benjamin and Cynthia Chou, pp. 7–76. Leiden: International Institute for Asian Studies and Singapore: Institute of Southeast Asian Studies.

Biesele, Megan. 1993. *Women like meat: The folklore and foraging ideology of the Kalahari Ju/'hoan*. Bloomington: Indiana University Press.

Bloch, Maurice. 1995. "People into places: Zafimanry concepts of clarity." In *The anthropology of landscape: Perspectives on place and space*, edited by Eric Hirsch and Michael O. Hanlon, pp. 63–77. Oxford: Clarendon.

Blust, Robert. 1981. Linguistic evidence for some early Austronesian taboos. *American Anthropologist* 83:285–319.

Borie, H. 1887. "An account of the Mantras, a savage tribe in the Malay Peninsula." In *Miscellaneous papers relating to Indo-China and the Indian archipelago*, pp. 286–307. Reprinted for the Straits Branch of the Royal Asiatic Society, second series, vol. 1. London: Trübner and Hill, Ludgate Hill.

Brookfield, Harold, Lesley Potter, and Yvonne Byron. 1997. *In place of the forest: Environmental and socio-economic transformation in Borneo and the eastern Malay Peninsular*. New York: United Nations.

Brosius, J. Peter. 1986. River, forest and mountain: The Penan Gang landscape. *Sarawak Museum Journal* 36:173–84.

———. 1991. Foraging in tropical rain forests: The case of the Penan of Sarawak, East Malaysia (Borneo). *Human Ecology* 19:123–50.

———. 1997. Prior transcripts, divergent paths: Resistance and acquiescence to logging in Sarawak. *Comparative Studies of Society and History* 39:468–510.

Burenhult, Niclas. 2002. A grammar of Jahai. Doctoral dissertation, Lund University.

Caldecott, Julian O. 1986. *An ecological and behavioral study of the pig-tailed macaque.* Contributions to Primatology vol. 21. Basil: Karger.

Campbell, Alan Tormaid. 1995. *Getting to know Waiwai: An Amazonian ethnography.* London and New York: Routledge.

Cant, R. G. 1973. *An historical geography of Pahang.* Kuala Lumpur: Monograph no. 4, Malaysian Branch of the Royal Asiatic Society.

Carey, Iskandar. 1976. *Orang Asli: The aboriginal tribes of Peninsular Malaysia.* Kuala Lumpur: Oxford University Press.

Casey, Edward S. 1996. "How to get from space to place in a fairly short stretch of time: Phenomenological prolegomena." In *Senses of place*, edited by Steven Feld and Keith H. Basso, pp. 13–52. Sante Fe, NM: School of American Research.

Certeau, Michel de. 1984. *The practice of everyday life.* Berkeley: University of California Press.

Chapman, F. Spencer. 1997. *The jungle is neutral.* Singapore: Times Book International. Original ed. 1949.

Christensen, Hanne. 2002. *Ethnobotany of the Iban and the Kelabit.* A joint publication of Forest Department Sarawak, Malaysia; NEPCon, Denmark; and University of Aarhus, Denmark.

Clifford, Hugh. 1992. *Report of an expedition into Trengganu and Kelantan in 1895.* Kuala Lumpur: Monograph no. 13, Malaysian Branch of the Royal Asiatic Society. Original ed. 1961.

Conklin, Beth A., and Laura R. Graham. 1995. The shifting middle ground: Amazonian Indians and eco-politics. *American Anthropologist* 97:695–710.

Cooke, Fadzilah Majid. 1999. *The challenge of sustainable forests: Forest resource policy in Malaysia, 1970–1995.* St. Leonards, NSW, Australia: Allen & Unwin and Honolulu: University of Hawaii Press.

Cronon, William. 1996. "The trouble with wilderness; or, getting back to the wrong nature." In *Uncommon ground: Rethinking the human place in nature*, edited by William Cronon, pp. 69–90. New York: Norton.

de Jong, W., U. Chokkalingam, J. Smith, and C. Sabogal. 2001. "Tropical secondary forests in Asia: Introduction and synthesis." In *Secondary forests in Asia: Their diversity, importance, and role in future environmental management* (=*Journal of Tropical Forest Science* vol. 13[4]), edited by U. Chokkalingam, W. de Jong, J. Smith, and C. Sabogal, pp. 563–76. Kepong: Forest Research Institute of Malaysia.

Dentan, Robert K. 1967. The mammalian taxonomy of the Sen'oi Semai. *Malayan Nature Journal* 20:100–106.

———. 1975. "If there were no Malays, who would the Semai be." In *Pluralism in Malaysia: Myth and reality* (=*Contributions to South East Asian ethnography*, vol. 7), edited by Judith Nagata, pp. 50–64.

———. 1979. *The Semai: A non-violent people of Malaya.* New York: Holt, Rinehart & Winston. Fieldwork ed.

———. 1991. Potential food sources for foragers in Malaysian rainforest: Sagos, yams and lots of little things. *Bijdragen Tot de Taal-, Land-en Volkendunde* 147:420–44.

———. 1997. "The persistence of received truth: How ruling class Malays construct Orang Asli identity." In *Indigenous peoples and the state: Politics, land, and ethnicity in the Malayan Peninsula and Borneo*, edited by Robert L. Winzeler, pp. 98–134. New Haven, CT: Yale Southeast Asia Studies 46, Yale University Press.

———. 2002a. "Against the kingdom of the beast: Semai theology, pre-Aryan religion and the dynamics of abjection." In *Tribal communities in the Malay world: Historical, social and cultural perspectives*, edited by Geoffrey Benjamin and Cynthia Chou, pp. 206–236. Leiden: IIAS and Singapore: ISEAS.

———. 2002b. 'Disreputable magicians,' the Dark Destroyer, and the Trickster Lord: Reflections on Semai religion and a possible common religious base in South and Southeast Asia. *Asian Anthropology* 1:153–94.

Dentan, Robert K., Kirk Endicott, Alberto G. Gomes, and M. B. Hooker. 1997. *Malaysia and the original people: A case study of the impact of development on indigenous peoples*. Needham Heights, MA: Allyn and Bacon.

Descola, Philippe. 1994. *In the society of nature: A native ecology in Amazonia*. Cambridge: Cambridge University Press.

Diamond, Jared. 1991. "Interview techniques in ethnobiology." In *Man and a half: Essays in Pacific anthropology and ethnobiology in honour of Ralph Bulmer*, edited by Andrew Pawley, pp. 83–86. Auckland: The Polynesian Society.

Douglas, Mary. 1966. *Purity and danger: An analysis of the concepts of pollution and taboo*. London and Henley: Routledge and Kegan Paul.

Dove, Michael R. 1983. Theories of swidden agriculture and the political economy of ignorance. *Agroforestry Systems* 1:85–99.

———. 1992. The dialectical history of jungle in Pakistan: An examination of the relationship between nature and culture. *Journal of Anthropological Research* 48:231–53.

———. 1996. Rice-eating rubber and people-eating Governments: Peasant versus state critiques of rubber development in colonial Borneo. *Ethnohistory* 43:33–63.

———. 1998. Living rubber, dead land, and persisting systems in Borneo: Indigenous representations of sustainability. *Bijdragen Tot de Taal-, Land- en Volkenkunde* 154:20–54.

(Father) Dunn. 1992. "The creation." In *The Sea Dyaks and other races of Sarawak: Contributions to the Sarawak Gazette between 1888 and 1930*, edited by Anthony Richards, pp. 27–29. Kuala Lumpur: Dewan Bahasa dan Pustaka. Original ed. 1963.

Dunn, Frederick L. 1975. *Rainforest collectors and traders: A study of resource utilization in modern and ancient Malaya*. Kuala Lumpur: Monograph no. 5, Malayan Branch of the Royal Asiatic Society. Reprinted 1982.

Ellen, Roy. 1999. "Forest knowledge, forest transformation: Political contingency, historical ecology and the renegotiation of nature in Central Seram." In *Transforming the Indonesian uplands: Marginality, power and production*, edited by Tania Murray Li, pp. 131–57. Amsterdam: Harwood Academic Publishers.

Endicott, Karen L. 1979. Batek Negrito sex roles. M.A. thesis, Australian National University.

———. 1981. The conditions of egalitarian male-female relationships in foraging societies. *Canberra Anthropology* 4:1–10.

———. 1992. "Fathering in an egalitarian society." In *Father-child relations: Cultural and biosocial contexts*, edited by Barry S. Hewlett, pp. 281–96. New York: Aldine de Gruyter.

Endicott, Kirk. 1970. *An analysis of Malay magic*. Singapore: Oxford University Press.

———. 1974. Batek Negrito economy and social organization. Ph.D. dissertation, Harvard University.

———. 1979a. *Batek Negrito religion: The world-view and rituals of a hunting and gathering people of Peninsular Malaysia*. Oxford: Clarendon.

———. 1979b. The hunting methods of the Batek Negritos of Malaysia: A problem of alternatives. *Canberra Anthropology* 2:7–22.

————. 1983. "The effects of slave raiding on the aborigines of the Malay Peninsula." In *Slavery, bondage, and dependency in Southeast Asia*, edited by Anthony Reid and J. Brewster, pp. 216–45. Brisbane, Australia: University of Queensland Press.

————. 1984. The economy of the Batek of Malaysia: Annual and historical perspectives. *Research in Economic Anthropology* 6:29–52.

————. 1995. "Seasonal variations in the foraging economy and camp size of the Batek of Malaysia." In *Dimensions of tradition and development in Malaysia*, edited by Rokiah Talib and Tan Chee Beng, pp. 239–54. Petaling Jaya, Malaysia: Pelanduk.

————. 1997. "Batek history, interethnic relations, and subgroup dynamics." In *Indigenous peoples and the state: Politics, land, and ethnicity in the Malayan Peninsula and Borneo*, edited by Robert L. Winzeler, pp. 30–50. New Haven, CT: Yale Southeast Asia Studies 46, Yale University Press.

Endicott, Kirk, and Peter Bellwood. 1991. The possibility of independent foraging in the rain forest of Peninsular Malaysia. *Human Ecology* 19:151–85.

Evans, Ivor H. N. 1923. *Studies in religion, folk-lore, and custom in British North Borneo and the Malay Peninsula*. Cambridge: The University Press.

————. 1937. *The Negritos of Malaya*. Cambridge: The University Press.

Fairhead, James, and Melissa Leach. 1996. *Misreading the African landscape: Society and ecology in a forest-savanna mosaic*. Cambridge: Cambridge University Press.

Feld, Steven, and Keith H. Basso, eds. 1996. *Senses of place*. Sante Fe, NM: School of American Research.

Flint, Kate, and Howard Morphy, eds. 2000. *Culture, landscape, and the environment: The Linacre Lectures 1997*. Oxford: Oxford University Press.

Foo; Eng Lee. 1972. *The ethnobotany of the Orang Asli, Malaysia, with special reference to their foodcrops*. Edited by Tan Koonlin. Unpublished typescript. University of Malaya School of Biological Sciences, Botany Unit, Kuala Lumpur.

Geertz, Hildred. 1989. *The Javanese family: A study of kinship and socialization*. Prospect Heights, IL: Waveland. Original ed. 1961.

Gell, Alfred. 1999. "The language of the forest: Landscape and phonological iconism in Umeda." In *The art of anthropology: Essays and diagrams*, pp. 232–58. London: Athlone Press.

Gomez, Edmund Terence, and K. S. Jomo. 1997. *Malaysia's political economy: Politics, patronage and profits*. Cambridge: Cambridge University Press.

Greenough, Paul. 2001. "*Naturae ferae*: Wild animals in South Asia and the standard environmental narrative." In *Agrarian studies: Synthetic work at the cutting edge*, edited by James C. Scott and Nina Bhatt, pp. 141–85. New Delhi: Oxford University Press.

Guemple, Lee. 1988. "Teaching social relations to Inuit children." In *Hunters and gatherers, vol. 2, Property, power and ideology*, edited by Tim Ingold, David Riches, and James Woodburn, pp. 131–49. Oxford: Berg.

Guenther, Mathias. 1988. "Animals in Bushman thought, myth and art." In *Hunters and gatherers, vol. 2, Property, power and ideology*, edited by Tim Ingold, David Riches, and James Woodburn, pp. 192–202. Oxford: Berg.

Gullick, J. M. 1988. *Indigenous political systems of Western Malaya*. London: Athlone.

Harris, David R. 1996. "Domesticatory relationships of people, plants and animals." In *Redefining nature: Ecology, culture and domestication*, edited by Roy F. Ellen and Katsuyoshi Fukui, pp. 437–63. Oxford: Berg.

Hirsch, Eric, and Michael O' Hanlon, eds. 1995. *The anthropology of landscape: Perspectives on space and place*. Oxford: Clarendon.

Hood Salleh. 1990. "Orang Asli of Malaysia: An overview of recent development policy and its impact." In *Tribal peoples and development in Southeast Asia [Special issue of the journal Manusia dan Masyarakat]*, edited by Lim Teck Ghee and Alberto G. Gomes, pp. 141–49. Kuala Lumpur: Department of Anthropology and Sociology, University of Malaya.

Hoskins, Janet, ed. 1996. *Headhunting and the social imagination in Southeast Asia*. Stanford, CA: Stanford University Press.

Howell, Signe. 1989a. *Society and cosmos: Chewong of Peninsular Malaysia*. Chicago: University of Chicago Press.

———. 1989b. "To be angry is not to be human, but to be fearful is: Chewong concepts of human nature." In *Societies at peace*, edited by Signe Howell and Roy Willis, pp. 45–59. London: Routledge.

Hutchins, Edwin. 1995. *Cognition in the wild*. Cambridge, MA: MIT Press.

Hutterer, Karl. 1985. "People and nature in the tropics: Remarks concerning ecological relationships." In *Cultural values and human ecology in Southeast Asia*, edited by Karl L. Hutterer, A. Terry Rambo, and George Lovelace, pp. 55–76. Ann Arbor, MI: University of Michigan Center for South and Southeast Asian Studies.

Ichikawa, Mitsuo. 1992. Comment on: "Beyond the original affluent society: A culturalist reformulation?" by Nurit Bird-David. *Current Anthropology* 33:40–41.

———. 1998. The birds as indicators of the invisible world: Ethno-ornithology of the Mbuti hunter-gatherers. *African Study Monographs* Suppl. 25:105–21.

Ingold, Tim. 1996. "Hunting and gathering as ways of perceiving the environment." In *Redefining nature: Ecology, culture and domestication*, edited by Roy F. Ellen and Katsuyoshi Fukui, pp. 117–55. Oxford: Berg.

Ismail, Rose. 1995. We must do more for the Orang Asli. *New Sunday Times* June 25.

Jabatan Hal-Ehwal Orang Asli (JHEOA), Malaysia. 1961. *Statement of policy regarding the administration of the Orang Asli of Peninsular Malaysia*. Department of Orang Asli Affairs.

Jomo, K. S. 1992. "The continuing pillage of Sarawak's forests." In *Logging against the Natives of Sarawak*, pp. v–ix. Kuala Lumpur: INSAN. 2nd ed.

Jones, Alun. 1968. The Orang Asli: An outline of their progress in modern Malaya. *Journal of Southeast Asian History* 9:286–305.

Kathirithamby-Wells, Jeya. 1997. "Human impact on large mammal populations in Peninsular Malaysia from the nineteenth to the mid-twentieth century." In *Paper landscapes: Explorations in the environmental history of Indonesia*, edited by Peter Boomgaard, Freek Colombijn, and David Henley, pp. 215–41. Leiden: KITLV Press.

Kato, Tsuyoshi. 1991. When rubber came: The Negeri Sembilan experience. *Southeast Asian Studies* 29:109–57.

Kawada, Junzo. 1996. "Human dimensions in the sound universe." In *Redefining nature: Ecology, culture and domestication*, edited by Roy F. Ellen and Katsuyoshi Fukui, pp. 39–60. Oxford: Berg.

Kelsall, H. J. 1894. Account of a trip up the Pahang, Tembeling, and Tahan Rivers, and an attempt to reach Gunong Tahan. *Journal of the Straits Branch of the Royal Asiatic Society* 25:33–56.

Kent, Susan. 1989. "Cross-cultural perceptions of farmers as hunters and the value of meat." In *Farmers as hunters: The implications of sedentism*, edited by Susan Kent, pp. 1–17. Cambridge: The University Press.

Keyser, Arthur. 1993. "Cuddling up to a tiger." In *They came to Malaya: A traveller's anthology*, edited by J. M. Gullick, pp. 271–72. Kuala Lumpur: Oxford in Asia.

Kitchener, H. J. 1961. "The importance of protecting the Malayan tiger." In *Nature conservation in western Malaysia, 1961*, edited by J. Wyatt-Smith and P.R. Wycherley, pp. 202–6. Malayan Nature Journal 21st Anniversary, Special Issue. Kuala Lumpur: Malayan Nature Society.

Knight, John, ed. 2000. *Natural enemies: People-wildlife conflicts in anthropological perspective*. London and New York: Routledge.

Kua Kia Soong, ed. 2001. *People before profits: The rights of Malaysian communities in development*. Petaling Jaya, Selangor: Strategic Info Research Development and Suara Rakyat Malaysia.

Kuchikura, Yukio. 1987. *Subsistence ecology among Semoq Beri hunter-gatherers of Peninsular Malaysia*. Sapporo, Japan: Hokkaido University Dept. of Behavioral Science.

———. 1996. "Fishing in the tropical rain forest: Utilization of aquatic resources among the Semaq Beri hunter-gatherers of Peninsular Malaysia." In *Coastal foragers in transition*, vol. 42, edited by Tomoya Akimichi, pp. 147–74. Osaka, Japan: Senri Ethnological Studies, National Museum of Ethnology.

Laidlaw, F. F. 1953. Travels in Kelantan, Terengganu and upper Perak, a personal narrative. *Journal of the Malayan Branch of the Royal Asiatic Society* 26:148–64.

Lakoff, George, and Mark Johnson. 1980. *Metaphors we live by*. Chicago: Chicago University Press.

Levi-Strauss, Claude. 1966. *The savage mind*. Chicago: Chicago University Press.

Li, Tania Murray. 1999. "Marginality, power and production: Analyzing upland transformations." In *Transforming the Indonesian uplands: Marginality, power and production*, edited by Tania Murray Li, pp. 1–44. Amsterdam: Harwood Academic Publishers.

Locke, A. 1993. *The tigers of Trengganu*. Kuala Lumpur: Monograph no. 23, Malaysian Branch of the Royal Asiatic Society. Original ed. 1954.

Lye Tuck-Po. 1994. Batek hep: Culture, nature, and the folklore of a Malaysian forest people. M.A. thesis, University of Hawai'i at Manoa.

———. 1997. Knowledge, forest, and hunter-gatherer movement: The Batek of Pahang, Malaysia. Ph.D. dissertation, University of Hawai'i at Manoa.

———, ed. 2001. *Orang Asli of Peninsular Malaysia: A comprehensive and annotated bibliography*. CSEAS Research Report Series no. 88. Kyoto: Centre for Southeast Asian Studies, Kyoto University.

———. 2002. "Forest peoples, conservation boundaries, and the problem of 'modernity' in Malaysia." In *Tribal communities in the Malay world: Historical, cultural and social perspectives*, edited by Geoffrey Benjamin and Cynthia Chou, pp. 160–84. Leiden: IIAS and Singapore: ISEAS.

———. 2004. "Uneasy bedfellows? Contrasting models of biodiversity maintenance in Malaysia." In *Biodiversity and society in Southeast Asia: Case studies of the interface between nature and culture*, edited by Michael R. Dove, Percy S. Sajise, and Amity Doolittle. New Haven: Yale Southeast Asia Council Press.

Maxwell, George. 1982. *In Malay forests*. Singapore: Eastern Universities Press. Original ed. 1907.

Meilleur, Brien A. 1994. "In search of 'keystone societies.'" In *Eating on the wild side: The pharmacologic, ecologic, and social implications of using noncultigens*, edited by Nina L. Etkin, pp. 259–80. Tucson, AZ: University of Arizona Press.

Miklucho-Maclay, Nicolas. 1878. Ethnological excursions in the Malay Peninsula. *Journal of the Straits Branch of the Royal Asiatic Society* 2:205–21.

Needham, Rodney. 1967. "Blood, thunder, and the mockery of animals." In *Myth and cosmos: Readings in mythology and symbolism*, edited by John Middleton, pp. 271–86. Austin, TX: American Museum Sourcebooks in Anthropology, University of Texas Press.

———. 1976. Minor reports concerning Negritos in northern Pahang. *Journal of the Malaysian Branch of the Royal Asiatic Society* 49:184–93.

Nelson, Richard K. 1983. *Make prayers to the raven: A Koyukon view of the Northern Forest*. Chicago: Chicago University Press.

Nicholas, Colin. 2000. *The Orang Asli and the contest for resources: Indigenous politics, development and identity in Peninsular Malaysia*. Copenhagen: IWGIA Document no. 95, International Work Group for Indigenous Affairs and Kuala Lumpur: Center for Orang Asli Concerns.

Nishimoto, Yoichi. 1998. Northern Thai Christian Lahu narratives of inferiority: A study of social experience. M.A. thesis, Chiang Mai University.

Noone, H. D. 1936. Report on the settlements and welfare of the Ple-Temiar Senoi of the Perak-Kelantan watershed. *Journal of the Federated Malay States Museums* 19 (Part 1):1–85.

Ohnuki-Tierney, Emiko. 1993. *Rice as self: Japanese identities through time*. Princeton, NJ: Princeton University Press.

Ooi Jin-Bee. 1963. *Land, people and economy in Malaya*. London: Longmans.

Peluso, Nancy Lee. 1996. Fruit trees and family trees in an anthropogenic forest: Ethics of access, property zones, and environmental change in Indonesia. *Comparative Studies of Society and History* 38:510–48.

Pillay, M. S. 1996. "Health and the environment." In *State of the environment in Malaysia*, edited by Consumers' Association of Penang, pp. 432–37. Penang, Malaysia: Consumers' Association of Penang.

Puri, Rajindra K. 1997. Hunting knowledge of the Penan Benalui. Ph.D. dissertation, University of Hawai'i at Manoa.

Rai, Navin K. 1985. "Ecology in ideology: An example from the Agta foragers." In *The Agta of Northeastern Luzon: Recent studies*, edited by P. Bion Griffin and Agnes Estioko-Griffin, pp. 33-44. Philippines: University of San Carlos.

Rambo, A. Terry. 1978. Bows, blowpipes and blunderbusses: Ecological implications of weapons change among the Malaysian Negritos. *The Malaysian Nature Journal* 32:209–16.

———. 1979. Primitive man's impact on genetic resources of the Malaysian tropical rain forest. *Malaysian Applied Biology* 8:59–65.

———. 1982. "Orang Asli adaptive strategies: Implications for Malaysian natural resource development planning." In *Too rapid rural development: Perceptions and perspectives from Southeast Asia*, edited by Colin MacAndrews and Chia Lin Sen, pp. 251–99. Ohio: Ohio University Press.

Razha, Rashid. 1973. "The Kintak-Bong of Tasek, Ulu Perak." In *Three studies on the Orang Asli in Ulu Perak*, edited by Shuichi Nagata, pp. 1–38. Penang, Malaysia: Perpustakaan Universiti Sains Malaysia.

Read, Peter. 2000. *Belonging: Australians, place and aboriginal ownership*. Cambridge: Cambridge University Press.

Ridley, H. N. 1893. On the dispersal of seeds by mammals. *Journal of the Royal Asiatic Society Singapore Branch* 25:11–32.

Rival, Laura M., ed. 1998. *The social life of trees: Anthropological perspectives on tree symbolism*. Oxford: Berg.

Roseman, Marina. 1991. *Healing sounds from the Malaysian rainforest: Temiar music and medicine*. California: University of California Press.

————. 1998. Temiar singers of the landscape: Song, history, and property rights in the Malaysian rain forest. *American Anthropologist* 100:106–21.

Ruslan; bin Said. 1990/1991. Konversi agama di kalangan masyarakat Orang Asli: Satu kajian di perkampungan Orang Asli Kampung Lebir Gua Musang Kelantan. Graduation exercise, Universiti Kebangsaan Malaysia [in Malay: Religious conversion among Orang Asli societies: A study in the Orang Asli settlements of Kampung Lebir, Gua Musang, Kelantan].

Sahabat Alam Malaysia. 2001. *Malaysian environment: Alert 2001*. Pulau Pinang: Sahabat Alam Malaysia.

Schama, Simon. 1990. *Landscape and memory*. New York: Knopf.

Schebesta, Paul. 1926. The jungle tribes of the Malay Peninsula. *Bulletin of the School of Oriental and African Studies* 4:269–78.

————. 1973. *Among the forest dwarfs of Malaya*. London: Oxford University Press. Reprint of 1928 ed.

Schefold, Reimar. 2002. "Visions of the wilderness on Siberut in a comparative Southeast Asian perspective." In *Tribal communities in the Malay world: Historical, cultural and social perspectives*, edited by Geoffrey Benjamin and Cynthia Chou, pp. 422–38. Leiden: IIAS and Singapore: ISEAS.

Scott, James. 1985. *Weapons of the weak: Everyday forms of peasant resistance*. New Haven, CT: Yale University Press.

Sellato, Bernard. 1993. Myth, history and modern cultural identity among hunter-gatherers: A Borneo case. *Journal of Southeast Asian Studies* 24:18–43.

Serpell, James. 1996. *In the company of animals: A study of human-animal relationships*. Cambridge: Canto edition of Cambridge University Press.

Shore, Bradd. 1996. *Culture in mind: Cognition, culture, and the problem of meaning*. Oxford: Oxford University Press.

Skeat, Walter William. 1901. *Fables and folk-tales from an eastern forest*. Cambridge: Cambridge University Press.

————. 1953. Reminiscences of the Cambridge University expedition to the north-eastern Malay states, 1899–1900. *Journal of the Malaysian Branch of the Royal Asiatic Society* 26:9–147.

————, and Charles O. Blagden. 1906a. *Pagan races of the Malay Peninsula, vol. 1*. reprinted 1966. London: MacMillan.

————, and Charles O. Blagden. 1906b. *Pagan races of the Malay Peninsula, vol. 2*. reprinted 1966. London: MacMillan.

Skinner, A. M. 1878. Geography of the Malay Peninsula, part 1. *Journal of the Straits Branch of the Royal Asiatic Society* 1:52–62.

Song, S. Hoon. 2000. "The Great Pigeon Massacre in a deindustrializing American region." In *Natural enemies: People-wildlife conflicts in anthropological perspective*, edited by John Knight, pp. 212–28. London and New York: Routledge.

Taussig, Michael T. 1980. *The devil and commodity fetishism in South America*. Chapel Hill, NC: University of North Carolina Press.

Teh Tiong-Sa, Voon Phin-Keong, Chan Kok-Eng, Tan Wan-Hin, and Tan Lee Seng. 2001. "Geography of Malaysia." In *Malaysia: National response strategies to climate change*, edited by Chong Ah Look and Philip Mathews, pp. 203–46. Kuala Lumpur: Ministry of Science, Technology and the Environment, Malaysia.

Tenas, Effendy. 2002. "The Orang Petalangan of Riau and their forest environment." In *Tribal communities in the Malay world: Historical, cultural and social perspectives*, edited by Geoffrey Benjamin and Cynthia Chou, pp. 364–83. Leiden: IIAS and Singapore: ISEAS.

Testart, Alain. 1982. The significance of food storage among hunter-gatherers: Residence patterns, population densities, and social inequalities. *Current Anthropology* 23:523–37.

Tsing, Anna Lowenhaupt. 1993. *In the realm of the diamond queen: Marginality in an out-of-the-way place.* Princeton, NJ: Princeton University Press.

Verheij, E. W. M., and R. E. Coronel, eds. 1992. *Plant resources of South-East Asia No. 2: Edible fruits and nuts.* Bogor, Indonesia: PROSEA.

Vogt, Christian. 1995. A friend of the Batek. *Aliran Monthly* no. 15:30–33.

Wagner, Roy. 1986. *Symbols that stand for themselves.* Chicago: Chicago University Press.

Wazir-Jahan, Karim. 1981. *Ma' Betisék concepts of living things.* London: Athlone.

Weiner, James F. 1991. *The empty place: Poetry, space and being among the Foi of Papua New Guinea.* Bloomington: Indiana University Press.

Wells, Carveth. 1925. *Six years in the Malayan jungle.* Singapore: Oxford in Asia Paperbacks, Oxford University Press.

Wessing, Robert. 1986. *The soul of ambiguity: The tiger in Southeast Asia*: Special report no. 24, Center for Southeast Asian Studies, Northern Illinois University.

Wharton, C. H. 1968. Man, fire and wild cattle in Southeast Asia. *Annual Proceedings of the Tall Timbers Fire Ecology Conference* 8:107–67.

Whitmore, T. C. 1997. *An introduction to tropical rain forests.* Oxford: Clarendon. Rev. ed.

Widlok, Thomas. 1997. Orientation in the wild: The shared cognition of Hai//om bushpeople. *Journal of the Royal Anthropological Institute* 3:317–32.

Wildlife Commission of Malaya. 1932. *Report of the Wild Life Commission.* 3 vols. Singapore: Singapore Government Printing Office.

Woodburn, James. 1982. Egalitarian societies. *Man* 17:431–51.

———. 1997. Indigenous discrimination: The ideological basis for local discrimination against hunter-gatherer minorities in Sub-Saharan Africa. *Ethnic Studies and Racial Studies* 20 no. 2 (April):345–51.

Yen, Douglas E. 1989. "The domestication of environment." In *Foraging and farming: The evolution of plant exploitation,* edited by David R. Harris and Gordon C. Hillman. London: Unwin Hyman.

Zawawi Ibrahim, ed. 1996a. *Kami bukan anti-pembangunan (Bicara Orang Asli menuju Wawasan 2020).* Kuala Lumpur: Persuatan Sains Social Malaysia. [in Malay: *We are not anti-development: Orang Asli discussion on Vision 2020*]

———. 1996b. The making of a subaltern discourse in the Malaysian nation-state: New subjectivities and the poetics of Orang Asli dispossession and identity. *Southeast Asian Studies* 34:568–600.

Index

affect. *See ha?ip*

affinities between Batek and the broader world, 15–17, 19, 28–29, 41–42, 170, 178

agriculture, 13, 70–71, 83–84, 89–90, 97, 128, 173. *See also* domestication of the environment; oil palm; plantation estates

alternatives and choices, 36, 38–39, 41, 170, 172–73, 175–76, 177

ambivalence. *See* forest

animal/people relations, 61–62, 72, 85–87, 111–17, 133–34, 140, 142–43. *See also* food; hunting; myths and folklore; sharing; wildlife conservation

animals: knowledge and observations of, 59, 60, 114; origins of, 80, 83, 85–86, 90; population decline of, 112, 115–16, 123; taxonomy of, 60–61, 90. *See also* birds; elephants; macaques; tigers

approach of the study, 2, 17, 40, 42–46, 169–71, 174

assimilation. *See* Department of Orang Asli Affairs

audience for the Batek, 15, 19, 26, 28–29, 170, 173, 176, 177. *See also* decision making

autonomy, 13, 51, 105–6, 173

Batek Nòng, xx–xxi, 5, 15

Batek Tanum, xx–xxi, 5, 167, 168

birds: appeal of, 61, 151–52; as indicators, 152–53; names of, 152; and pollination, 86. *See also* mimicry; myths and folklore

Blɛ̃?, 167

boundary status, 169, 177–78. *See also* foragers of ideas

broad spectrum foraging, 62–63

Bumekel Stream (Tɔm Bumǝkǝl), 97

camps: composition, 11, 12; as cultural symbol, 54, 68–69, 158; distances

between, 158; geography and locations of, 1, 6, 15, 24, 54, 55, 56, 57, 58, 77, 136–37, 167; names of, 69; and noise, 155–56; and pathways, 6, 43, 66, 67; perspectives from, 50, 69–70, 71; population numbers of, 12, 24, 25, 141–42; as sites of social memory, 99, 158; spatial distinctions in, 71–72; and tiger infestation, 116, 150; threats to, 22, 72–73, 97–98. *See also* knowledge; lean-tos; movement; pathways; place

change: directions of, 172–73; in ideologies, 50, 73, 78–79, 88–89, 92, 117–19, 144, 169, 171, 172–73, 176; rejection of, 39, 41, 170; social change, 2, 6–7. *See also* cultural persistence; landscape change

Chewong, 59, 107, 120n36

children: animal interactions of, 62, 114, 153, 154; births and deaths of, 97, 100, 167, 168; disciplining of, 108, 154, 158; foraging of, 13, 64, 124, 126, 137–38, 139, 154; interaction with Malays, 108; noises of, 156; sentiments of, 99; and slavery, 103

collecting forest products, 10, 13, 77, 89, 104, 108, 128, 135

colonialism, 14, 73, 74n36, 104, 106–8; colonial visitors and explorers, 95, 103, 147

communication: with the broader world, 1–2, 21–23, 28–29, 31, 34, 36, 38–39, 40–42, 46, 169–71, 173, 177; intergroup, 11, 67, 116, 137. *See also* audience for the Batek; decision making; hearing; sounds; Tebu

conquest narratives, 104

consciousness raising. *See* knowledge

cultural persistence, 78, 91, 123–24, 171, 172–73. *See also* change

dam building, 10, 22, 38, 118, 168, 170, 173

About the Author

Lye Tuck-Po is an independent anthropologist based in Kuala Kubu Baru, Malaysia. She received her graduate degrees at the University of Hawai'i at Manoa, where she was also a Degree Fellow at the East-West Center. Previous positions include the Quillian Visiting International Professorship at Randolph-Macon Woman's College in Virginia and fellowships at the Center for Southeast Asian Studies, Kyoto University, and the Research School of Asian and Pacific Studies, Australian National University. She has also worked as a newspaper reporter, conservation consultant, and climate change activist. One of her long-term interests is how to integrate local resource management practices with scientific conservation. She edited *Orang Asli of Peninsular Malaysia: A Comprehensive and Annotated Bibliography* (Center for Southeast Asian Studies, Kyoto University, 2001) and, with Wil de Jong and Abe Ken-ichi, *The Political Ecology of Tropical Forests in Southeast Asia: Historical Perspectives* (Kyoto University Press and TransPacific Press, 2003). This is her first book on the Batek.

- rambling prose, oral quality, conversational
- away from h/g as living fossils   tacks back &
                                         forth quickly
                                         btwn subjects

142: cultivated vs. wild.
      - a number of interesting detailed ex. & Qs.
      = haunted by notion of intentionality

149: soundscape